Cameroon's Tycoon

Cameroon Studies

General Editors: *Shirley Ardener, E.M. Chilver* and *Ian Fowler*, Queen Elizabeth House, International Development Centre, Univeristy of Oxford

Volume 1
Kingdom on Mount Cameroon. Studies in the History of the Cameroon Coast, 1500–1970 – Edwin Ardener. Edited and with an Introduction by Shirley Ardener.

Volume 2
African Crossroads. Intersections between History and Anthropology in Cameroon – Edited by Ian Fowler and David Zeitlyn

Volume 3
Cameroon's Tycoon. Max Esser's Expedition and its Consequences – Edited by E.M. Chilver and Ute Röschenthaler

CAMEROON'S TYCOON

Max Esser's Expedition and its Consequences

Edited by E.M. Chilver and Ute Röschenthaler

Berghahn Books
New York • Oxford

First published in 2001 by **Berghahn Books**

www.berghahnbooks.com

© 2001 E.M. Chilver and Ute Röschenthaler

All rights reserved.
No part of this publication may be reproduced in any form or by any means without the written permission of Berghahn Books.

Library of Congress Cataloging-in-Publication Data

Esser, Max.
 [An der Westküste Afrikas. English. Selections]
 Cameroon's tycoon : Max Esser's expedition and its consequences / edited by E.M. Chilver and Ute Röschenthaler.
   p. cm. -- (Cameroon studies : v. 3)
 Includes bibliographical references (p.) and index.
 ISBN 1-57181-988-6 (acid-free paper) -- ISBN 1-57181-310-1 (pbk. : acid-free paper)
   1. Cameroon--History. 2. Africa, West--Description and travel. 3. Plantations--Cameroon--History. 4. Cameroon--Economic conditions--To 1960. 5. Bali (African people) 6. Esser, Max. 7. Germans--Cameroon--History. I. Chilver, E. M. II. Röschenthaler, Ute. III. Title. IV. Series.

DT574.E8713 2001
967.11'02'092--dc21   2001035397

British Library Cataloguing in Publication Data
A catalogue record for this book is available from the British Library.

Printed in the United States on acid-free paper

ISBN 1-57181-988-6 hardback
1-57181-310-1 paperback

Max Esser in his late middle age: a Studio portrait, reproduced by kind permission of the Prinzessin Ratibor. This portrait was used with others in Wilhelm Kemner's pamphlet, *Was wir verloren haben,* 1922.

TABLE OF CONTENTS

List of Maps and Illustrations	ix
Abbreviations	xii
Preface by the Editors	xiii
Editorial Note	xvi
Acknowledgements	xviii

PART I THE SETTING, PUBLIC AND PRIVATE

1.	Max Esser: His Life and Labours by *Ute Röschenthaler*	3
	Key to Family Photographs	28

PART II ESSER'S TRAVELS

2.	The Outward Voyage	31
3.	Sao Thomé and Principe	35
4.	Cameroon – the Historical Background	43
5.	Land and People in Cameroon	49
6.	In Cameroon	57
7.	The Expedition to Bali	65
8.	Departure from Cameroon	113
9.	Angola, and the Cunene Expedition	117
10.	A Retrospective View	125

PART III COLONIAL NEEDS AND THEIR CONSEQUENCES: THE VIEWPOINTS OF SOME CONTEMPORARY OBSERVERS

11.	The 'Bali Road' and Baliburg in the Autumn of 1892: a Report on a Visit: Max von Stetten	133

12. A Complication: the Entry of the *Gesellschaft Nordwest Kamerun*, 1901–1903: Esser's Correspondence 141

13. A Parliamentary Visitation: Johannes Semler's Togo und Kamerun: Eindrücke und Momentaryfuahmen vou einem dentschen Abgeordneten, Leipzig, 1905 147

14. A Soldier's View of the Tasks of the Bamenda Military Station 1908: Hptm. Menzel 153

15. Labour Supply: a Shift of Modalities, 1913: Hptm. Adametz 159

Appendix I   The 'Esser Affair' 165

Appendix II   The 'Fetishes' and the Esser Collection at the Linden Museum, by *Ute Röschenthaler* 171

Maps 185

Select Bibliography 189

Index 201

List of Maps and Illustrations

Maps

Map 1: Esser's route map

Map 2: A language map

Illustrations

Frontispiece: Max Esser in his late middle age: a Studio portrait, reproduced by kind permission of the Prinzessin Ratibor. This portrait was used with others in Wilhelm Kemner's pamphlet, *Was wir verloren haben,* 1922.

Figure 1.1: The Luxemburg connection: from the photograph album of Max Esser's daughter Elisabeth, made available by courtesy of Herr Josef Pfaffenlehner, Schloss Sandersdorf.

Figure 1.2: A Christmas Party: the Essers and the Ahns
(By kind permission of the Prinzessin Ratibor)

Figure 1.3: Max Esser and his daughter Elisabeth
(By kind permission of the Prinzessin Ratibor)

Figure 1.4: Robert Esser's 80th birthday in 1913: a family gathering in Baden-Baden
(By kind permission of the Prinzessin Ratibor)

Figure 1.5: Number 6 Ahlsenstrasse, Berlin Mitte, Esser's residence between 1906 and 1917. The tiny figures on the upper balcony are those of his daughter and her governess, Mademoiselle Mion.
(From the photograph album of Max Esser's daughter)

Figure 1.6: The villa in Baden-Baden
(By kind permission of the Gräfin zu Eulenburg)

Figure 1.7: From the middle years of Max Esser (All from his daughter's photograph album, made available by courtesy of Herr Josef Pfaffenlehner, Schloss Sandersdorf.)

Figure 2.1: Part of the ruins of Great-Fredericksburg fort near Takoradi: from Wilhelm Kemner's photograph album.

Figure 6.1: The much-abused steamship *Nachtigal*

Figure 7.1: Zintgraff in expedition costume, from the frontispiece to *Nord-Kamerun* (see chapter 6, note 1)

Figure 7.2: The 'Fetishes' collected by Esser in Ikiliwindi (the tallest) and Bulo Nguti, as illustrated in Esser's book.

Figure 7.3: Mfon Galega of Bali-Nyonga – a photograph by Zintgraff reproduced in the *Koloniale Zeitschrift*, 1900.

Figure 7.4: Bali-Nyonga seen from Baliburg – a photograph illustrating an article by Franz Hutter in the *Koloniale Zeitschrift*, 1900.

Figure 10.1: The young Max Esser with Victor Hoesch and others in Cameroon: an early undated print from his daughter's photograph album, which was made available by courtesy of Herr Josef Pfaffenlehner, Schloss Sandersdorf.

Figure 10.2: WAPV beginnings in 1897 at Kakaohafen: from Kemner's pamphlet of 1922

Figure 13.1: Penal labourers (J. Semler, 1905)

Figure 13.2: The Schloss at Buea (J. Semler, 1905)

Figure 14.1: The proclamation of Fonyonga II as Paramount Chief on 15 June, 1905, in the presence of Captain Glauning, Lt. von Puttlitz, the Basel Missionaries Ernst and Keller, soldiers and delegations from 47 other Chiefdoms: photograph by Missionary Goering. (Reproduced by kind permission of the Basel Mission: photo archive No. K. 776)

Figure 14.2: Bali Soldiers (*Basoge*) in a motley of German uniforms: c.1907–8 (Ankermann Collection, No.315 D – Reproduced by courtesy of the Museum für Völkerkunde, Berlin)

Figure 15.1 A lasting connection: a visit by Kemner and members of the WAPV management to Fonyonga II in the 1930's (From Wilhelm Kemner's photograph album).

Figure AII.1: Royal Wives in Bali-Nyonga, 1907-9, one carrying a spear (Ankermann Collection No. 310. Reproduced by courtesy of the Museum für Völkerkunde, Berlin)

Figure AII.2: Two of the famous *nggwasi*, 'Women's garments' (10942, 10943) (By courtesy of the Linden Museum Stuttgart)

Figure AII.3: A 'hairpin' (hair ornament), 'fetish' (medicine) horn, three metal bangles, and a necklace made of brass cartridge cases (10935, 3682, 3263, 3658, 3662, 3698), Items from the Esser Collection at the Linden Museum.(By courtesy of the Linden Museum)

Figure AII.4: The three 'fetishes' from Bulo Nguti presented by Esser to the Linden Museum (3395, 3394, 3393)
(By courtesy of the Linden Museum).

Abbrevations

ABM – Archives of the Basel Mission Basel
ANY – Archives nationales Yaounde
BAB – Bundesarchiv Berlin-Lichterfelde
DKB – Deutsches Kolonialblatt
DKZ – Deutsche Kolonialzeitung
GNK – Gesellschaft Nordwest Kamerun
GSK – Gesellschaft Süd-Kamerun
JASO – Journal of the Anthropological Society of Oxford
KLPG – Kameruner Land- und Plantagengesellschaft
MDS – Mitteilungen aus den deutschen Schutzgebieten
WAPV – Westafrikanische Pflanzungsgesellschaft Victoria

Preface by the Editors

In March 1898 the *Deutsche Kolonialzeitung*, the organ of the German Colonial Society, informed its readers of the publication of a new book in the following terms:

'In the Spring of 1896 Dr. Esser set off, in the company of two other gentlemen, Dr. Zintgraff and Herr Victor Hoesch, to investigate whether, and which, particular enterprises would have a prospect of profitable success in our colonies and to indulge in a bit of hunting on their way. The enthusiasm of the late Dr. Zintgraff had clearly provided the motive power for the expedition; Zintgraff had worked tirelessly for many years to develop the Protectorate and lived to see his efforts crowned with success. The travellers visited Principe, Sao Thomé and Cameroon, and our readers will certainly be struck by Dr. Esser's lively report on Cameroon and everyday life in Bali . . . The travellers went on to Loanda, explored the little-known southern hinterland of the Portuguese colony and returned along the right bank of the Cunene river to Mossamedes via Tiger and Alexander Bays. Their explorations have since been picked up, from another point of view, and completed by Professor Dr. Wohltmann and have inspired new plantation enterprises along the right lines. The book is written in a sprightly style and is handsomely produced.'

A century has now passed since the publication of Max Esser's *An der Westküste Afrikas* which he sub-titled *Wirtschaftliche und Jagd-Streifzüge* and which might be translated as 'Business and Hunting Forays'. His book describes the early stages of an important development in Cameroon's economic history, that of plantation enterprise on a grand scale and the labour migration it gave rise to. In the scramble for Africa a new imperial Germany had quite recently acquired a large land-mass with a barely explored hinterland sandwiched between areas under French and British influence. Neither the men in the governing institutions of the Second *Reich* nor its thriving business community had yet developed very clear ideas about what was to be done with it and which were its most promising fields of investment. It was a good moment for reassessment of the economic opportunities it offered by an independent entrepreneur with good connections.

In 1896 Max Esser went to Cameroon for the first time. A young merchant banker, with a taste for adventure and hunting, he re-explored the prospects for a large-scale plantation industry around Mt. Cameroon: its people had been finally subdued in 1894. He then travelled the 'Bali Road' to persuade the *Mfon,* the ruler of Bali-Nyonga, to send labourers down to the coast. This was the first of several extended visits to Cameroon.

Esser's book is clearly intended to arouse more enthusiasm in its readers for Germany's acquisition; past and present colonial aspirations and an imperial destiny are described with fervour. It can also be counted among the earlier sources for the historical ethnography of the hinterland of the coast of the present South-West Province and the edge of the Western grassfields, coming as it does between Eugen Zintgraff's *Nord-Kamerun* of 1895 and Franz Hutter's *Nord-Hinterland* of 1902.

Apart from first-hand exploration and campaign accounts by Esser's contemporaries, published in official and learned journals, there are memoirs, a large array of specialist, especially geographical, literature and unofficial nostalgic accounts by colonists. After the Second World War the Colonial past was again re-examined, both from an ethnographic viewpoint and from a more critical, often Marxist or neo-Marxist one, in particular as the wealth of the reassembled colonial archives began to be appreciated. At present the roots, or rather, rhizomes, to borrow J-F. Bayart's metaphor, of the post-colonial, multi-ethnic state are again attracting the political economists as capitalism is re-invented and African scholars become increasingly involved in the interpretation of the past in the present.

But we shall not be attempting an analysis of past writings, still less of the amply-documented contests of the historians, but return the reader to the beginnings of an important development in the person of one of its pioneers, whose appearance on the Cameroon scene coincided with metropolitan situations conducive to speculative overseas investment, given a bit of luck and the right connections. We give Esser's account of it and of some of the consequences as they appeared to contemporary observers in the fifteen years following the inception of his grand plantation plan.

There were longer-term consequences. One was the creation of a modernised agro-industrial sector in isolation from Nigeria on the one hand, and relatively remote from the later centre of political power and patronage in Yaounde on the other. The social and demographic effects of labour migration can be followed in E.W. Ardener's publications (some conveniently collected in Volume 1 of this series). For another we could point to the cost and difficulty of maintaining what can still be viewed as the 'Road to Bali', an old problem casting a backward shadow, crossing, as it does, corrugated forest country to climb a steep escarpment, at least for as long as the 'Bali lands' remained a major source of labour and other easier routes were not developed.

Editorial Note

This book is divided into three parts, and Appendices. Part I is an essay by Ute Röschenthaler in which Max Esser's family connections and life history are explored, to the extent it has been possible to recover them.

In Part II we have presented a literal translation of that part of Esser's book which deals with his expedition to Bali-Nyonga, his agreement with its ruler to supply labour to his projected plantation company and his return journey.

Parts of his other chapters are summarized. Summarized parts have been indicated by square brackets. Why did we not translate this 225-page work entire? The summarized parts are often not of direct interest to Cameroon Studies, including his studies of the Portuguese islands of Principe and Sao Thomé and his travels in Angola on behalf of his backers. Some of them, too, consist of repetitive descriptions and reflexions which, in our opinion, added little or no information.

The Notes refer to Esser's sources of information, contemporary accounts and opinions, later assessments, some biographical details of other actors and later ethnographic references and identifications.

Part III consists of five short annotated translated texts – contemporary documents from archival or old published sources dealing with the crucial question of labour supply from the hinterland, starting with the state of the 'Bali Road' in the last days of Zintgraff's expedition and ending with the situation as seen by the Station Commander of the Bamenda Military Station in 1913. The 'opening up' of the hinterland also opened a Pandora's box of questions, including those of competition between incompatible German interests. The rendering of toponyms and ethnonyms in earlier reports presents quite a few puzzles, not only because of different realizations by contemporary interpreters but because of misreadings of hand-written reports and the likelihood of occasional typographical errors. Our own identifications are shown in square brackets.

Finally we have two Appendices. The first deals with the 'Esser Affair' touched on in Röschenthaler's essay and some of its reverberations. The second deals with the trophies Esser brought back from his travels and presented to the Linden Museum, Stuttgart, then in process of formation. Some quite intriguing issues arise.

Our Select Bibliography includes a few of the bibliographical works we have found helpful.

Acknowledgements

We have to thank very many kind helpers who forwarded our research, have allowed us the use of photographs, have put us on to useful sources of information, have shared their own unpublished information and memories, and met our insatiable appetite for photocopies of archival and early published material from newspapers, colonial and missionary journals. We list them below with renewed and heartfelt thanks.

Hermann Ahn, Köln: Dr. Cathérine Baroin, CNRS, Nanterre: Herr Bätje, Stadt Nordeney: Dr. Thomas Becker, Archivleitung, Universität Bonn: Ulrich Becker, Institut für Hochschulkunde, Würzburg: Thomas Begerow, Der Herold e. V., Berlin: Gregor Berghausen, Rheinisch-Westfälisches Wirtschaftsarchiv zu Köln: Jens Blecher, Universitätsarchiv, Leipzig: Ms. Brackmann, W.h. Bundesarchiv, Berlin-Lichterfelde: Dr. Anne Brandtstetter, Universität Mainz: Herr Braunn, Hauptstaatsarchiv Stuttgart: Beatrice di Brizio, Paris: Ms. Büttner, Staatsarchiv der Freien Hansestadt Hamburg: N. Davidson, Q.C.: Ms. Deissler, Bestattungswesen der Stadt Baden-Baden: Robert Ehrhard, Stadtarchiv Baden-Baden: Chief E. M. A. Epie, Kurume: Helmut Ernst, Wintershall AG, Kassel: Dr. Werner Esser, Köln: Heinz Christian Esser, Köln: Dr. Carl-Heinrich Esser, Mannheim: Heinrich Esser, Köln: Dr. Michael Esser, Baden-Baden: Gräfin zu Eulenburg, München: Dr. Hermann Forkl, Linden-Museum Stuttgart: Dr. Liesl Franzheim, Köln: Dr. C. Geary, Washington: Prof. Dr. Groten, Historisches Archiv, Köln: Dr. Reiner Haehling von Lanzenauer, Baden-Baden: Dr. Axel Harneit-Sievers, Centre for Modern Oriental Studies, Berlin: Herr Hertweck, Standesamt Baden-Baden: Rolf Heitlinger, Ettlin Gesellschaft für Spinnerei und Weberei AG, Ettlingen: Cathrin Hennicke, Leipzig: Doris Herdin, Berlin: Esther Hertlein, Würzburg: Herr Homburger, Amt für öffentliche Ordnung, Baden-Baden: Hans Walter von

Hülsen, Gernsbach: Dr. Paul Jenkins, Basel Mission, Basel: Dr. Peter Junge, Übersee-Museum Bremen: Dr. Georg von Kern, Bayerisches Armeemuseum, Ingolstadt: Dr. Hans-Joachim Koloss, Museum für Völkerkunde, Berlin: Rolf Kracke-Berndorff, Pulheim bei Köln: Ms Barbara Krupp, Esslingen: Per Lauke, Ahn und Simrock, Bühnen-und Musikverlag GmbH: Mairie de la Ville de La Celle Saint-Cloud, Service de l'Etat Civil, La Celle St.-Cloud: Map Room, Royal Geographical Society: Professor T. Mbuagbaw, Buea University: Mr. Menzel, Stadt Guben: Andreas Merz, Basel Mission, Basel: Petra Metzger, Stattreisen Köln: Vanessa Meynen, Zürich: Josef Pfaffenlehner, Schloss Sandersdorf: Dr. Manfred Pohl, Historisches Institut, Deutsche Bank, Frankfurt: Edith Przyrembel, Westdeutsche Gesellschaft für Familienkunde e. V., Brühl: Ms.A. Raab-Rebentisch, Historisches Institut, Deutsche Bank, Frankfurt: Prinzessin Ratibor, Schloss Unterriexingen: Dr. Ritter, W.h. Bundesarchiv, Berlin-Lichterfelde: Sieghardt Schaupp, Evangelisches Pfarramt der Markus-Gemeinde Baden-Baden: Andrea Schneider, Gesellschaft für Unternehmensgeschichte e.V. Frankfurt: Ms. Schneider, Securitas Versicherungen, Bremen: Edith Seidel-Sichter, Berlin: Dr. Christine Seige, Museum für Völkerkunde, Leipzig: Dr. Heinz Spangenberg, Traditionsverband ehemaliger Schutz- und Überseetruppen, München: Ms. Spauschus, Grünflächenamt der Stadt Köln: Ms. Sprenger, Köln: Christine Stelzig, Museum für W.h. Völkerkunde, Berlin: Holger Stöcker, Berlin: Bob Townsend, Queen Elizabeth House Library, Oxford: Ms Zandeck, Militärarchiv (Bundesarchiv), Freiburg: Henri Zuber, Archives Nationales, Centre historique de Paris.

The Copyright illustrations derived from the photo-archives of the Linden Museum, Stuttgart, the Basel Mission, and the Museum für Völkerkunde, Berlin are reproduced with their kind permission.

U. Röschenthaler would like to express her thanks for the many kindnesses shown and the helpful information given to her by chiefs, regents and community leaders in the course of a return trip from Kembong to Buea, via Tali, Nguti, Manyemen, Talangaye, Kokobuma, Konye, Kurume, Baduma, Ikiliwindi, Mambanda, Kumba, Mundame, Mukonye and Malende: Nfon E.M.E. Eseme VI of Kokobuma; Chief Manfred Mbende Bolloh of Mbakwa Supe; Nfon Muki Simon Esika of Kombone; Chief Collins Mbele Ekale of Ikiliwindi; Mfon William Mbo Aduma of Mambanda; Madam Theresa Fomenki and Chief Joseph Ndongo Ewanne II of Mundame; Nfon E.M.A. Epie of Kurume; Mr. Bate Mbu Ete of Mboka; Mr. Mbongya Tomas Fobi of Ekenge; Chief Mbuk Martin

Eben of Manyemen; Chief Enaw Benson Ebai of Talangaye; Chief Andreas Akame of Babensi I; Chief A. Fobia IV traditional ruler of Nguti, and his wife.

Special thanks are due to Shirley Ardener, who found time to keep the Editors, one in Berlin, one in Oxford, in touch by e-mail despite many other pressing tasks, and found for us, in Buea University, colleagues, among whom we are especially grateful to Dr. E. S. D. Fomin, who could unravel some of the ethnographic puzzles presented by Esser's route. And finally to Yvonne Gardiner who patiently made sense of some unusually difficult copy in three languages.

Part I

The Setting, Public and Private

CHAPTER 1

Max Esser: His Life and Labours

Ute Röschenthaler

The answer to the question 'What do we know about Max Esser' is all too little. His book, published in a fine illustrated edition by his brother-in-law Albert Ahn, provides few clues to his life and background. His comparisons between the costumes he devises to please the ruler of Bali with those of the Cologne carnival might identify him as a Rhinelander. From his book, we can also pick up the fact that he had learned musketry drill during his military service.

We know of Esser's leading role in the establishment of large-scale cocoa and other plantations on the coastal mountain range of Cameroon from colonial memoirs, for example, those of Jesco von Puttkamer (1912), Theodor Seitz (1927/29) and Wilhelm Kemner (1922, 1937/41), from the federal German archives and from company reports. Kemner, Esser's successor as managing director of the main plantation enterprise he created, the *Westafrikanische Pflanzungsgesellschaft* 'Victoria' (WAPV), describes him as a bold, far-sighted and energetic man, and explicitly acknowledges his venturesomeness as an investor. The Cameroonian economist Dr Simon Epale (1985), looking back to the forerunners of the Cameroon Development Corporation, refers to his able direction.

Max Esser was almost certainly the most noteworthy capitalist who appeared in person on the Cameroon scene during the German period. Historians of colonialism see Esser in quite another light, as a man whose activity could not fail to enserf or impoverish the peoples of the mountain by occupying so much of their land, and whose search for a cheaper local work force inevitably gave rise to some of the inhumanities of forced labour. This we can see, for example, in the works of Stoecker (1960, 1968), Ballhaus (1968), Hausen (1970) and Clarence-Smith (1993). The question of labour supply remained a cardinal issue for early plantation management, and frequently involved coercion at first or second hand, amid growing competition for labour due to the increasing demands by the colonial state, particularly by the railways, the trading companies and even the missions.

Some Family History

Max Esser, christened Maximilian Richard Karl Ludwig, was born on 14 September 1866, in Cologne, the second son of Robert Esser, a prominent company lawyer. The German equivalent of *Who's Who* and other genealogical directories make no mention of him in connection with this well-known Cologne legal family, as they do of his elder brother. The surname is a common one in the Rhine Province.[1]

We know next to nothing about his early life except that he obtained a doctorate in law (*D. jur.*) and that he was a Lieutenant (Cavalry) in the Reserve, the *Landwehr*. We do know that 'law student, Max Esser, from Cologne, Catholic' was registered in the Faculty of Law at the University of Freiburg in the summer semester of 1888. He must then have spent some time at Berlin University for he arrived from there at the University of Bonn where he spent three semesters, from October 1889 to March 1891, attending courses in the Faculty of Law. The Dean of the Faculty of Philosophy also signs his attendance record. Bonn University records indicate that when he had completed his studies Esser returned to Cologne. Its old university had been closed by the French in 1798 and had not yet reopened. So where did he complete his doctorate in law? We learn from the manuscript ledger of doctoral promotions in the University of Leipzig that he presented his thesis there, as an external student, *cum laude*, on 16 February 1891, over a month before he was formally exmatriculated from Bonn, on 27th March. His chosen topic was: 'The conflict between Aquilian and strict liability in tort, as well as the practical effects arising therefrom', referring to the Roman *lex Aquilia* (named after an early Roman jurist), and thus to the Roman Dutch Civil Law tort system which is based on that law. No copy of the thesis has been traced nor does it figure in bibliographies among works issuing from German-speaking Universities. This could well be accounted for by the fact that, to judge from its 1892 regulations, the University of Leipzig did not oblige doctoral candidates to present a given number of copies of their thesis in printed form, unlike other Universities. (Berlin University, for example, explicitly required such candidates to provide 250 copies, printed at their own expense). The examiners' copy could well have been lost with other such student materials during and after the Second World War (Jarausch, 1982, 428). We shall probably never know why Esser chose to acquire his doctorate at Leipzig University.

In 1894, at the age of 28, he appears for the first time in Berlin directories as resident at Schiffbauerdamm 6, in the centre of the city, facing the River Spree. At that time he was employed in managing the Berlin office of the *A. Schaaffhausen'scher Bankverein*, a Rhineland merchant bank of renown, spoken of in the same breath as the *Darmstädter Bank* and *Leipziger Credit-Anstalt*, which had opened a Berlin branch in 1891.

A closer look at his family background will help to explain how much Max Esser owed to his family's connections, professional contacts and expertise, all needed for his forthcoming travels in Cameroon and the Portuguese colonies.

His paternal grandfather, Ferdinand Joseph Esser (1802–1871) was a lawyer in Cologne and *Geheimer Justizrat*, a title broadly corresponding to that of a British Q.C. As president of the *Zentral-Dombau-Verein* he was concerned with the preservation of historic sites and the promotion of the arts. In 1829 he married another Esser, Karoline (1810–1883), the eldest daughter of Ignaz Joseph Esser, scion of another prominent Cologne family, that of the State Councillor (*Regierungsrat*) Reiner Esser (1747–1833). The only amply recorded son of their marriage was Max Esser's father, Robert Joseph Esser (1833–1920).

Robert Joseph Esser also became a lawyer, was awarded an honorary doctorate, became a *Geheimer Justizrat* and was otherwise honoured by the Prussian state. He was awarded the *Roter Adlerorden*, the *Kronenorden* 2nd Class with Star, and other decorations. He owned two houses in Cologne.[2] He was a leading expert in company law on which he wrote a number of books. By 1913 he was Chairman of nine companies, including the *A. Schaaffhausen'scher Bankverein* of Cologne, and the *Rheinisch-Westfälische Bodenkreditbank*, and a board member of thirteen other companies, among them the *Dresdner Bank*, the Phoenix Insurance group, and the Gelsenkirchner Mining Company.[3] He was briefly (from 1898–1900) a member of the board of the *Gesellschaft Süd-Kamerun* (GSK), a monopoly concession company initiated by, among others, the influential Hamburg financier Dr Julius Scharlach (1842–1908) and the mining industry magnate Sholto von Douglas (1837–1912). These two men were involved in many other overseas enterprises, including some in the Congo and South Africa (Ballhaus, 1968, 105, 110).

Abraham Schaaffhausen's Bank[4] was converted in 1848 under Prussian law from a private to a public company and Robert J. Esser was a member of the board until his death. It was this bank that was later used for most of Max Esser's Cameroon plantation transactions. Some of the family's marriages with other banking families are not without interest. Max Esser's future mother-in-law was born a Deichmann,[5] an important banking family in Cologne, which had owned Abraham Schaaffhausen's Bank before its conversion, and Max Esser's great-uncle, Karoline Esser's brother, had married Therèse Mertens-Schaaffhausen.

Like his father before him, Robert Esser was involved in various bodies for the promotion of the arts; for example, as president of the *Zentral-Dombau-Verein* and a member of both the municipal arts advisory board and the management committee of the local museum.[6]

In 1864 he had married Adele Josephine von Kaufmann (1845–1919), a leading Cologne figure in her own right. She was, for example, a member of the committee of management of a women's patriotic organization, the *Deutschnationaler Frauenverein*. By 1883 she had become chairwoman of the Cologne branch of the *Deutscher Frauenverein für Krankenpflege in den Kolonien,* a precursor of the Colonial Red Cross. She headed a fund-raising body for daughters of officers killed in the Franco-Prussian war. She founded and was, for many years, in charge of a convalescent home for nursing sisters, and was a Dame of the *Luisenorden,* a Prussian State decoration. Adele was also awarded the Red Cross Medal, 2nd Class, and other honours (Steimel, 1958). She was the daughter of an ennobled and baptized convert from Judaism, Jakob von Kaufmann-Asser (1819–1875), who had made a considerable fortune in commodity trading (*Briefadel Taschenbuch*, 1917) and had later founded one of the new-style investment banks (*Effekten-Banken*)[7] in Cologne. Early in its life the 1873 depression supervened, and Max Esser's maternal grandfather, unable to meet his obligations, is said to have committed suicide by drowning himself in the Rhine. Or so alleged the anti-Semitic *Deutsch-Soziale Blätter* (a sheet which had a pretext for later reviving the tale in its issues of 17 and 31 August 1899).

Robert and Adele Esser had four children – two sons, Ferdinand (1865–1927) and Max (1866–1943), and two daughters, Henriette (b.1869) and Karoline (b.1875).[8] Max's sister Henriette, in 1891, married the publisher Albert Ahn (*D. jur.*) of Cologne (1869–1935), a good friend to Max and his family, and the publisher of his book; they had four children: a son and three daughters.

Max's elder brother Ferdinand stayed on in Cologne and followed in his father's footsteps. He studied law (*D. jur.*), became a *Justizrat* and married Maria Kreuser (1871–1910) of Cologne in 1891. Their notable son, Robert Esser the younger (1895–1969) became proprietor and manager of the *Bankhaus Ferdinand Schroeder AG*, in Cologne, and vice-president of the Cologne Chamber of Commerce and Industry. He was also a member of the managing board of the Rhineland-Westphalia Stock Exchange at Düsseldorf.[9] Like his father and paternal grandfather, he was a committee member of the *Zentral-Dombau-Verein*. Also a Catholic, he married Annemarie Plate-Voigtsdorf and had two daughters.

The Cameroon Project

Max Esser's family background demonstrates that he belonged to the old-established *haute bourgeoisie* of Cologne. While we have no specific evidence to suggest an early shift of interest to colonial affairs on his own part, his parents' early interest in them has been amply recorded (Soénius, 1992), as well as that of the Cologne business community in general. In 1881 Friedrich Fabri, supervisor of the Rhenish Missionary Society and author of *Bedarf Deutschland der Kolonien?* (1879), organized with the support of other leading figures in Rhineland trade and industry, the *Westdeutscher Verein für Kolonisation und Export* (WVKE), of which Robert Esser the elder was an active promoter. This was followed in 1882 by the *Deutscher Kolonial-Verein* involving many of the same members, which merged with Carl Peters' Berlin organization in 1888 to form the principal colonial lobby, the *Deutsche Kolonialgesellschaft* (DKG). The DKG subsumed the varied interests of the existing groups in the acquisition of colonies as a source of raw materials for the German Empire's rapidly expanding industries, an outlet for emigration overseas, or an export market and a field for profitable investment, once rights in land and mining rights had been acquired on easy terms. The Cologne branch was an active one with a special interest in Cameroon, possibly further stimulated by Hugo Zöller's reports of his West African travels for the local newspaper, the *Kölnische Zeitung*, in 1884–5. It would be hard to believe that Max Esser had been left untouched by the colonial debates of his young manhood.

Shipping and trading interests apart, the initial investment in Cameroon plantations in 1885 by the Hamburg interests (represented by Woermann, and Jantzen and Thormählen), had not been followed up by a rush to acquire land and invest in its long-term development. Rather, it was railways that remained in fashion and attracted the lion's share of investment. In 1895 a consortium of bankers and industrialists, which included the influential Adolf von Hansemann of the *Disconto-Gesellschaft*, one of the major banks, the potash magnate Wilhelm Oechelhäuser and the sugar manufacturer Eugen Langen, had collected funds for an expedition to Angola. This was, apparently, to explore better means of access to the north of German South-West Africa and the possibility of building a railway from a southern port of Angola along its borderlands. The aim was to connect it with the Transvaal and ultimately, to reach an Indian Ocean port. The consortium entrusted the young banker Max Esser with organizing a further preliminary survey from the Portuguese side: a railway project had already been mooted by the Anglo-German South-West Africa Company with an eye to the possible acquisition of Tiger Bay harbour. According to Kemner (1922), Esser had made contact with the Portuguese Mossamedes concession company and had acquired a conditional option for the consortium on a stretch of land it possessed. Esser planned the expedition to southern Angola with his former fellow-student Victor Hoesch (who was to be a witness to his civil marriage). Hoesch is described as 'a gentleman of independent means' (*Rentner*) and came from a Düren factory-owning family. Hoesch introduced Esser to Eugen Zintgraff of Coburg, the first explorer to reach the Cameroon Grassfields, Adamawa and the Benue from the coast.

Zintgraff had made several official expeditions to the interior between 1886 and 1892, the last jointly with Jantzen and Thormählen traders. In 1889 he had been hospitably received in the chiefdom of Bali-Nyonga, which he made his headquarters for further exploration to Adamawa and the Benue waterway. His later attempts to open up trade with its northern and larger neighbour Bafut had been violently rebuffed: Zintgraff's moves to exact revenge for his murdered envoys had resulted in a military defeat in 1891 by Bafut and its allies, the loss of guns, and the deaths of four

Europeans. Reinforcements and modern arms had since arrived in Bali, but the then Governor, von Zimmerer, had been either unwilling or unable to support Zintgraff, distrusting his adventurism. The expedition was recalled in January 1893. Zintgraff's plan, to open up a route to the populous Grassfields – seen as a source of manpower for the police and the plantations – and to set up permanent stations and settled agricultural villages along it, was curtly turned down by the Foreign Office which refused to lift the Governor's ban on his return.[10]

Esser met Zintgraff when he was in this situation and it was Zintgraff, it seems, who persuaded Esser and Hoesch to include Cameroon in their itinerary. At the same time a new Governor, Jesco von Puttkamer, was on leave, and Esser reintroduced him to Zintgraff. The ban on Zintgraff's return to Cameroon was lifted.

Von Puttkamer had already been convinced by his friend Spengler, a Thuringian engineer, planter and German Vice-Consul of the Portuguese island-colony of Sao Thomé, of the suitability of the lands around Mt Cameroon for plantations. He was already in negotiation with the Foreign Office's colonial section to acquire a large tract of land to be sold on to interested investors. Esser and his backers fitted in perfectly with the Governor's plans. There remained however the question of labour supply which had already taxed the managers of Woermann's *Kameruner Land- und Plantagengesellschaft* (KLPG) and the Jantzen and Thormählen tobacco and cocoa-growing concern, both of which had depended to a large extent on expensively imported labourers. Zintgraff proposed a remedy: they should visit his old ally, the ruler (*Mfon*) of Bali-Nyonga, and involve him in the recruitment of labour for the plantation company they were planning to set up. Zintgraff's treaty of 1891 with this ruler, reported in the official *Kolonialblatt*, recognized him as 'the paramount chief of the surrounding tribes of the Northern Cameroon hinterland', established him as a German ally and agent, and promised him a subsidy.

According to von Puttkamer, Esser 'agreed on the spot' to take charge of a major plantation enterprise near Victoria (now Limbe) provided he could examine the site before organizing the necessary capital investment. Zintgraff had also suggested that they should take in Sao Thomé on their way so as to examine Spengler's successful plantation enterprise.

On May 6, 1896, after a merry farewell dinner at the Berlin Club (Puttkamer, 1912, 69) Esser, Hoesch and Zintgraff set out on a Portuguese steamer.

Following their visit to Spengler and the expedition to Bali-Nyonga – the central part of this book – Esser and Hoesch carried out their assignment in southern Angola, joined by a 'white hunter', Walter Fournier-Baudach, who had previously been to Angola. Esser and Hoesch returned to Berlin at the end of 1896. As Kemner puts it, they had left with 'a problematic plan for a railway and returned with a sizeable, safe, plantation enterprise.' Their proposals were favoured by investors, and the *Westafrikanische Pflanzungsgesellschaft* 'Victoria' (WAPV) was set up with an initial capital of 2.5 million Marks, a headquarters in Berlin and a branch office in Victoria. Esser was managing director, and Zintgraff (who died in Tenerife at the end of 1897) was its first plantation manager.[11]

Esser must have written his book in less than a year, despite other involvements. Nonetheless, it is a well-structured work with his visit to Bali and King Garega as its central point. Moreover, it is not lacking in other characteristics such as detailed and often poetic descriptions of the African natural world, the dense forest, and the montane Grassfields, which reminded him of the Black Forest and the Alps of Tyrol and Switzerland. It is also clear that the military spectacle put on by the Bali to impress their visitors struck a deep chord in him. The book can be seen as a sort of heroic prospectus, not only because it includes a good deal of useful information culled from many sources but also because of the optimism of its tone; problems did exist but could be surmounted with the necessary ingenuity.

At the time he wrote his book, Esser was living at Potsdamer Strasse 121, Berlin-Tiergarten, over the shop, as it were, as this address was also that of the headquarters of the WAPV. In the same year, 1897, Esser was involved with a group founding another plantation company, the *Westafrikanische Pflanzungsgesellschaft Bibundi* (WAPB), and was briefly a member of its board.[12] He went out to visit the WAPV enterprise in the autumn of 1897. In mid-1898 he found time to approach the colonial section of the Foreign Office with the Emperor's friend Sholto von Douglas and the Hamburg financier Julius Scharlach, to negotiate the setting up of the GSK concession company in the newly-explored lands of Cameroon's south-eastern hinterland. Although he withdrew from this scheme, which had attracted Belgian banks – his father, as we saw, briefly joined the board. In the autumn of 1898 we find Max back in Cameroon, energetically putting his enterprise in order. In the meantime he had found time to deliver lectures, including ones for the Berlin Geographical Society and the Colonial Society (DKG) to publish in their journals[13] and to prepare his book for publication.

The Esser Affair

In December 1898, on his return from Cameroon, Esser was given an audience by the Emperor and is said to have presented an exposition of the prospects of Cameroon that lasted an hour and a half, after which the Emperor engaged him in conversation, and awarded him the *Kronenorden* 2nd Class. He also favoured Esser with some complimentary words about his encouragement of private investment in the protectorate. The conversation was reported, possibly a trifle apocryphally, in the gossip press; Esser is supposed to have hinted at too much red tape in colonial administration. It seems likely enough that Esser was considered young, at 32, to have received such a high honour usually reserved for lengthy public service.[14]

The applause was soon followed by sharp criticism at the hands of one Dr Hans Wagner of Königsberg, correspondent of the Berlin *Tägliche Rundschau*, who openly accused Esser of deceiving the Berlin Geographical Society, the scientific community and worse, the Emperor, about the extent of his travels in southern Angola. Press attacks continued throughout 1899, and included a challenge to Esser to take his accuser to court for libel. Esser in return challenged him to a duel which was anonymously reported to the police and stopped. Esser was brought before the courts and condemned to three months' imprisonment, but the Emperor intervened to virtually quash the sentence. Thereafter a 'Court of Honour' busied itself with the question of whether Esser was fit to retain his commission.

The initial attack was taken up by the anti-Semitic press with some relish, and it dug out his maternal grandfather's suicide, accused Esser of financial dealings that had worried the stock-exchange's controlling body, and hinted that his private life was dubious. Wagner further accused him of ingratiating himself with the Imperial Court by gifts, and of using his undeniable eloquence and courtly manners in pursuit of his own financial interests as well as being the stalking-horse of unnamed financial interests.[15]

The anti-Semitic element of the campaign was the manifestation of a strong contemporary undercurrent. The Officers' Reserve, for example, had not only excluded Jewish officers in the 1880s on the grounds of their religion, but also later excluded those who were baptized Christians (Kitchen, 1968, 22–48 and John 1981, 153ff.). The anti-Semites also complained that the Cologne press had not publicized the 'Esser affair'. But the 'Esser affair' is not analogous to the Dreyfus affair. Esser, though well supported by both Victor Hoesch and Fournier-Baudach, took few pains to refute the allegation that he had not personally visited some of the places shown on his Angola itinerary. Neither did he dispute that some of his information was collected at second-hand. Disclaiming any map-making expertise for himself he stressed that he himself had concentrated on economic activities as indicated by his book's sub-title, *Business and Hunting Forays*. Although he was expelled from the Officers' Reserve he was allowed to keep his decoration. None of the accusations against him made by Wagner impugned Esser's Cameroon travels, although he was accused of liberally borrowing his scientific descriptions from the agronomist Wohltmann.

On May 15 of this troublesome year, Max Esser married Franziska Elisabeth (Lily) Dutreux (1871–1935), daughter of Antoine Auguste ('Tony') Dutreux, before the mayor of La Celle St. Cloud. Her father, a prominent Luxemburg deputy and industrialist with iron and steel interests, owned the local Château de la Celle St Cloud (Scheibler and Wülfrath, 1939).[16] She has been described to us as very elegant in her person and a good manager. It was thus that Emma Eleonore Flora Thérèse Dutreux, *née* Deichmann, became Max Esser's mother-in-law, and brought him close to one of the most important banking families of the period. As well as having been the proprietors of Schaaffhausen's private bank from 1830 until its transformation in 1848, they had been involved in the foundation of the International Bank of Luxemburg.[17] It was commonly said that Esser owed the greater part of his fortune to his wife; nevertheless he must have already been well-established as a man of means. He had one child by this marriage, a daughter, Elisabeth, born at La Celle, in 1900. She married twice. Her first marriage, later dissolved, was to Christian Jordan, a consular officer by whom she had a son, Tony, who died young, and a daughter, Johanna Maria, of whom we know next to nothing. Elisabeth's much happier second marriage was to Thomas, Baron de Bassus (1907–1989). She died in 1984 at Glion in Switzerland, to where the couple had finally moved after periods of residence in Bavaria and Morocco.

The plantation enterprises

Between 1899 and 1906 Max Esser was living at Sommerstrasse 6 in the centre of Berlin where the WAPV had moved its headquarters in July 1898. Together with his friend Victor Hoesch he made regular trips to Cameroon to supervise the plantations, usually for several months during the dry season. Apart from his visits in 1897 and 1898 he appears to have made such trips every eighteen months or so from 1900 up to 1908 (Kemner 1937: 137 and WAPV Annual Reports). Between 1899 and 1902, the greatest problem he encountered and strove to solve was the nutrition and health of the labour force for which the management's plans had proved totally inadequate, and had led to a shockingly high rate of mortality among labourers from the so-called Bali lands.

Esser's plantation interests expanded, despite problems with cocoa brown rot and other pests. He founded smaller plantations, one with Oechelhäuser, and had a light railway built between the port of Victoria and the WAPV plantations – Puttkamer mentions in his memoirs that he took a ride on it at the end of March 1905. His activities had even extended to oil prospecting; he set up experimental bore holes at Logobaba near Duala but was not granted the 10-year sole exploration licence he sought (Langheld, 1909: 418; von Puttkamer, 1912: 313–314). The company was liquidated and its installations sold for a song. Also in this period – between 1900 and 1907 – he held a position on the board of a Hanseatic company financing a German settlement in Brazil.

The headquarters office of the WAPV had moved to Unter den Linden, and from there another plantation company, the Debundscha Plantation Company, was organized in 1905 with Esser as its first chairman. He was already sometime chairman and board member of the Meanja Rubber Plantation (from 1903 to 1907) and in 1906 had been among the founders of the larger *Deutsche Kautschuk AG*, in which he had a large stake.[18] He was also a founding and board member of an insurance company, *Securitas Versicherungen AG*, and on the management boards of six mining companies. By April 1906 he had acquired his own separate residence at 6 Ahlsenstrasse in the Berlin-Tiergarten quarter near the river Spree.[19] In 1909 he acquired yet another house, Bismarckstrasse 4. According to Martin, in 1911 Esser had a fortune of 11 million Marks and an annual income of 0.8 millions, ranking as No. 52 of the millionaires in the Kingdom of Prussia.[20]

A setback in Esser's Cameroon ventures, insofar as the WAPV

was concerned (and not his other and smaller plantation interests), seems to have taken place after his last visit of inspection to Cameroon in 1907/8. For this we must turn to Kemner's account. Kemner, born in Cologne in 1872, had served in the army until 1900. Thereafter, he worked on the marketing side of a Lorraine mining company that fell into difficulties, and had to be reconstructed by the *A. Schaaffhausen'scher Bankverein*. In the course of this exercise he met Dr Ernst Schroeder who was shortly to succeed his father as Director of the Bank. Schroeder asked Kemner whether he would be interested in reconstructing a Cameroon plantation company, in which his bank held about 1 million Marks' worth of preference shares. Although the share capital had already been consolidated and new capital injected, the concern was still barely profitable. Kemner was invited to assist Esser, who subsequently introduced him to the WAPV Chairman, the great landowner Prince Alfred zu Löwenstein. The Prince engaged Kemner shortly before Esser set out on his last inspection. 'On his [Esser's] return,' writes Kemner, 'a difference of opinion arose between him and the Board which led to his resignation' (Kemner 1937: 137). Neither Kemner nor the surviving Board annual reports in the Berlin Federal Archives describe the nature of the disagreement. Esser's last report as managing director was in May 1908: it repeats a decision to abandon some cocoa and other plantings where they were not doing well in order to economize on running costs. Indeed, the fungal cocoa infestation, 'brown rot', had already seriously depleted harvests in the high rainfall areas, cocoa prices had fallen somewhat after a boom period, and the costs of technology transfer might have seemed unduly high and possibly overdone (Clarence-Smith, 1993: 203). However, none of this is specifically mentioned.

Kemner, who succeeded Esser as managing director claimed to have single-handedly turned round the WAPV, making it less vulnerable to fluctuations in the cocoa market, and more profitable by diversifying its production. These claims stand in need of some modification, however, for this process had already begun. Kemner met with a good deal of initial resistance from the Board in getting agreement to his plans.[21] These included a reallocation of local tasks and stricter financial control from Berlin.

What of Esser's other Cameroon interests? In 1908 the Meanja Rubber Plantation Company, set up with Esser as Chairman in 1903, underwent a reconstruction.[22] Plantation rubber was initially encouraged in Cameroon as an insurance in the longer term against the destructive methods of tapping wild rubber, collected and sold to trading and concession companies by weight. The native Kickxia was selected for propagation and the seed was sold to the coastal planters. When a new and abundant source of wild rubber was found interest in plantation rubber temporarily waned. But by 1904 the Government felt obliged to call a halt to the continued destruction of rubber producing plants. It was made a penal offence and native collectors were to be instructed in better methods. Kickxia continued to be planted but before it could show results it was discovered that Hevea was a better option in a highly competitive market. So plantation rubber had made a false start and it would be some eight years before any returns could be expected from the Hevea seeds distributed after 1906. This explains why so little plantation rubber was exported before 1914 (Epale, 1985: 36–37). It might also explain why, in 1908, Esser allowed a troublesome enterprise to be taken over by a group of investors led by H.F. Picht,[23] the energetic Chairman of the *Deutsche Kautschuk AG*, in 1908. Esser no longer remained on the Meanja Plantation Board, though his old friend Victor Hoesch and his brother Hermann Hoesch remained members of it. All three were major shareholders.

In 1910 Picht and some of his group gained control of the Board of the Debundscha Plantation Company. In 1911 Esser refused re-election to its Board. The Hoesch brothers remained. It is difficult to interpret these manoeuvres.

In 1910 Esser sold the house he had bought in the Bismarckstrasse a year after its acquisition. In 1916 he probably disposed of his house in the Ahlsenstrasse though he still appears in lists of residents, and is described as 'plantation-owner'. In 1917 and 1918 he was residing in the Bendlerstrasse,[24] still in the fashionable and attractive Tiergarten quarter, after which he ceased to be recorded as a resident in Berlin. At some point – we do not have a date for this – he acquired property in the fashionable health-resort of Baden-Baden. This was an impressive villa set in a handsome garden at 5 Ludwig-Wilhelm-Strasse. He was to fill it with antiques. It was in the resort of Baden-Baden, too, that the family had assembled to celebrate his father Robert Esser's eightieth birthday.

The photograph album of Max Esser's daughter provides us with some further glimpses of his life. In both 1912 and 1913 he went to the island sea-side resort of Norderney with his young daughter, her French governess and a maid. In 1913 we see him on the new racecourse at Norderney, in front of the bandstand, with Prince Adolf von Schaumburg-Lippe: Baden-Baden had its racecourse too. In the early years of the First World War, in 1915, we see him in full dress officer's uniform, at Guben, south-east of Berlin. In 1916, out of uniform and in his fiftieth year, we see him in conversation with a senior officer, Freiherr von Solemacher, in Brussels, which, of course, was then occupied by German forces. We do not know whether military use was made of his special knowledge: he remained on the boards of several companies, including the *Kali AG Berlin,* for example. Nor do we know of his activities during the turbulent years immediately after the war, which saw the occupation of the Ruhr by French and Belgian forces in 1922, and the galloping inflation of 1923. Despite these facts, as the photograph album shows, the Baden-Baden races were run as usual in 1923, attracting fashionable race-goers and cosmopolitan sporting figures.

Esser's later interests extended to insurance, shipping – the *Kameruner Schiffahrt* AG of Hamburg – and, nearer home, the Ettlin spinning and weaving works in Baden.[25] More importantly however, now that no restrictions were to be placed on the nationality of the purchasers, the former plantation companies were getting together to reacquire the Cameroon plantations at auctions in London scheduled for November 1924.

After the bulk of the plantations returned to German ownership following the auctions, Esser returned to the Board of the WAPV, still the largest plantation enterprise in Cameroon, and remained there from 1926 to 1936.[26]

We know that Esser died in Baden-Baden on February 6, 1943, aged 77,[27] following his wife who had died at Samaden in Switzerland in 1935. His family's impressive monumental graves are in the Melaten cemetery in Cologne – those of his parents Robert Esser and Adele von Kaufmann, that of his elder brother Ferdinand and his wife, and Ferdinand's son Robert and his wife. Not far from the Esser family graves is the vault of Albert Ahn and his family. Max Esser's remains are not there: the urn containing his ashes is buried in the public cemetery of Baden-Baden. His house there passed to his daughter and remained her property until at least 1950. By 1955 it had already changed hands.

So quite a few mysteries hang over the life and personality of Max Esser. How far, one wonders, was he playing to the anti-slavery gallery in his account of virtual slave labour in the Portuguese colonies? Or to the conservative land-owning interest in his assurances that they had nothing to fear from colonial expansion? It was inevitable that he viewed the Africans he met through the lenses of the popular Social Darwinism of the time, and sought to beguile them. However, some genuine enjoyment of their company, their ingenuity and their crafts is also apparent in his writing.[28]

Arguments about the economic value of smallholdings as opposed to plantations are as active as ever, and take several forms.[29] Esser can provide a convenient scapegoat for some critics. Others might regard him as the unconscious architect of incipient class formation and even inter-ethnic collaboration. We cannot, meanwhile, deny Esser's merits as a *raconteur* and as a delineator of the often contradictory sentiments aroused in those who took a leading part in exporting European capitalism to Cameroon.

Notes

1. The name 'Esser' can derive either from an old term for an axle-maker, or from a term for a member of a (medieval) Hundred, or just refer to a connection with arable land. (*Der Familienname Esser*, 1920.)

2. Robert Esser's residence, Am Hof 22, where Max grew up, was in the centre of Cologne, hard by the Cathedral. Destroyed in the Second World War, the site is now occupied by West German Radio. The family also owned a summer residence on the banks of the Rhine.

3. Martin, 1913, *Jahrbuch der Millionäre*, Vol.9, Rhine Province.

4. At first, the Prussian Government regarded the founding of private merchant banks (*Aktienbanken*) with some scepticism as speculative establishments rather than as the capital accumulating and investing bodies called for by industrialization. Rhenish bankers themselves, such as David Hansemann (Aachen), Abraham Oppenheim, Gustav Mevissen, and Ludolf Camphausen (all of Cologne) held different opinions. Finally, the Prussian Government decided that its Prime Minister Camphausen and Finance Minister Hansemann should be allowed to set up the *Schaaffhausenscher Bankverein* as the first of such banks to stabilize the situation; others followed (Pohl, 1993: 264, 269).

5. Wilhelm Ludwig Deichmann was a son-in-law of the Abraham Schaaffhausen who had founded the bank in 1791. Deichmann directed the bank from 1830 onwards (Steimel, *Kölner Köpfe*, 1958).

6. Information kindly supplied by Gregor Berghausen of the *Rheinisch-Westfälisches Wirtschaftsarchiv zu Köln*. The family's association with the *Zentral-Dombau-Verein* was not accidental. The completion of Cologne Cathedral, a patriotic project started in the 1840s, continued to elicit a sense of civic responsibility for the maintenance of the city's artistic heritage and was close to the heart of Kaiser Frederick William IV (H. James 1989: 47–48). Both Robert Esser and Ferdinand Joseph

(his father), as presidents of this body, were honoured by having their names and armorial bearings displayed in a cathedral window.

7. The number of *Effekten-Banken* (investment banks) mushroomed after the Franco-Prussian war as a consequence of more liberal banking regulations. In the 1873 depression there was a run on these banks, many of which were unable to meet their obligations (Pohl, 1993: 265).

8. Max's younger sister Karoline was married, in 1898, to Josef May (*D. jur.*) (1867–1899), who died shortly after; they had a daughter. In 1902 she married Georg von Caro (*D. jur.*), commercial councillor, who was 26 years older than she was. He was a landowner and a well-connected board member of several iron works and mining companies and the *Deutsche Kautschuk* AG. Among other honours he was awarded the *Kronenorden* 2nd class (Degener, 1912). Her third marriage was to Major Wolff, whom she survived.

9. Robert Esser, Max Esser's nephew, was the joint owner of the *Sauerstoffwerk Gewerkschaft Rose* in Obersteg, a member of the board of *AG Flora* and of the *Heinrich Stöcker AG* in Cologne (Stockhorst, 1967). He was also on the advisory or management boards of other local industries, a Cologne insurance company and the regional employers' organization for banking institutions (Degener, 12th edn., 1955).

10. For Zintgraff's explorations see Sally Chilver (1966, 1999) and Zintgraff's own account, *Nord-Kamerun*, 1895. Conrau (1898) describes the route a few years later on and tells us something of Zintgraff's personal agenda.

11. The first members of the board, according to a communication from Esser to Freiherr von Richthofen, head of the colonial section of the Foreign Office, in January 1897, were:
    Chairman: Prince Alfred zu Löwenstein-Wertheim-Freudenberg;
    Otto Andreae, a leading Cologne businessman, who became vice-chairman later, but meanwhile Sholto von Douglas deputized; Albert Ahn the publisher; the factory owner Max Hiller; Victor Hoesch; the State Councillor Professor von Kaufmann; the banker Carl Levy (later Hagen); the Düren financier Leopold Peill; Freiherr Julius von Soden, former Governor, and Dr Georg Seitz. A little later Morton von Douglas replaced Sholto. The major investors, in order of importance, were the Douglas group, Prince Alfred himself, the Esser-Ahn group of fifteen smaller investors – Esser himself took up one hundred and twenty five 1000-Mark shares –, the so-called *Deutsche Bank* group represented by Carl Levy, the Hoesch group, the Peill-Schöller group and the Hiller group. Düren was the site not only of the Hoesch works but of the Peill and Schöller sugar-refining business. Hugo Schöller was soon to join the board.

12. Other members included Victor Hoesch's brother Hermann, an iron and steel manufacturer, who later joined the WAPV board, the

diplomat von Kusserow, the influential Dr Julius Scharlach, member of the Colonial Council, W. Oechelhäuser, also a member and a leading industrialist, G.P. Dollmann, of Hamburg, already involved in the KLPG plantation, and Dr Ferdinand Wohltmann.

13. These were: 'Meine Reise nach dem Kunene im nördlichen Grenzgebiet von Deutsch-Südwest-Afrika', in *Verhandlungen der Gesellschaft für Erdkunde zu Berlin*, 24, 1897, 103–113, with map; 'Unsere Westafrikanischen Kolonien und ihr portugiesischer Nachbar', in *Beilage zur Deutschen Kolonialzeitung*, No. III, 6.2.1897, 9–15 (Pt 1), and No. VIII, 27.3.1897, 33–36 (Pt. 2). A shorter piece, 'Sitten der Hereros' appeared in the same journal, Vol. X, 1897, 193–194.

14. Esser was also awarded an Italian knightly order, the Crown of Italy, to be worn on the left breast, as we see in his photograph. The circumstances have escaped our search, but must have occurred before his marriage. The same applies to the *Ehrenkreuz* awarded to him by the small state of Schwarzburg-Sondershausen in Thuringia. (Information kindly supplied by Thomas Begerow, Herold e.V., Berlin)

15. Among the capitalists Wagner dubbed 'Scharfmacher', agitators, trying to 'persuade the Emperor that he would be deprived, as it were, of the capital invested in his work' he mentions Prince Alfred zu Löwenstein, first chairman of the WAPV board, and Prince Hohenlohe-Öhringen, Herzog von Ujest, promoter of the GNK concession company (*Tägliche Rundschau*, 6 October 1899).

16. The long-lived 'Tony' Dutreux (1838–1933) is described as a Commander of the Legion of Honour and as holding important Belgian and Luxemburg decorations in the 'acte' of civil marriage between Max Esser and Lily Dutreux. Apart from the Château de la Celle St. Cloud he owned land and a town-house in Luxemburg. His profession is given as engineer, trained at the famous *École Centrale* in Paris, and Vice-Chairman of an important iron and steelworks complex with the acronym ARBED. He also set up the *Fondation Pescatore*, named after his mother, in Luxemburg. Witnesses to the proceedings included Victor Hoesch and Robert Brasseur, a young lawyer living in Luxemburg, for Max Esser and, for his bride, a senior Belgian lawyer, her cousin, Octave Maus, and her brother, Auguste, an industrial engineer, who eventually became President of the Dunlop Tyre Company in France. (Records of the Commune of La Celle St. Cloud, 15 May 1899, fol. 22; Scheibler and Wülfrath, 1939, Vol. 1, under Deichmann).

17. We are grateful to Gregor Berghausen for this information. Wilhelm Ludwig Deichmann (1798–1876) not only directed the *A. Schaaffhausen'scher Bankverein* after its refounding, but started the *Bankhaus Deichmann & Co.*, and was a co-founder of the *Deutsche Bank AG* (Steimel, *Kölner Köpfe*, 1958). See also Martin, 1913: 187ff.

The Berlin branch of the *A. Schaaffhausen'scher Bankverein* started in 1891 was taken over by the *Disconto-Gesellschaft* in May 1914, while the Cologne and other branches still operated under the old name. In October 1929 there came about a major fusion between the *Deutsche Bank* and the *Disconto-Gesellschaft* whereby the name of the *A. Schaaffhausen'scher Bankverein* finally disappeared (Professor Dr Manfred Pohl, personal communication).

18. For Esser's activities on the Boards of plantation companies see company reports in the Federal Archives, Berlin-Lichterfelde (BAB): R1001/RKA 3499–3508, 3511–3514, 3529, 3534, 3539–3542, 3544. See also Martin 1913: 187ff.

19. The entire quarter of the Spreebogen was destroyed in the Second World War. New government buildings now occupy the site.

20. The Emperor Wilhelm II ranked as No. 5, after Bertha Krupp von Bohlen, Fürst Henkel von Donnersmarck, Freiherr von Goldschmidt-Rotschildt, and the Herzog von Ujest, but he had the highest income. In 1911 there were nearly 1,000 millionaires in Prussia. Martin's book, as he explains, is mainly based on the self-declarations of wealthy Prussians who had sent the author supplements and amendments to an earlier survey.

21. Shortly after his entry into the WAPV Kemner (1937: 137–8, and 1941: 40) undertook his first journey of orientation and inspection with Prince Alfred and Dr Hermann Hoesch, brother of Victor Hoesch. (See also BAB, R1001/RKA 3502) The company, he says, had been unprofitable until 1906. After Kemner's entry WAPV's share values had risen and rose rapidly until war broke out in 1914. For share and dividend values see Hausen (1970) 311, 315 and Kemner (1937) 164.

22. The reorganized Meanja Plantation Company was accused in the financial newspaper, *Graf's Finanzchronik*, of failing to mention considerable losses by the Company in the past, in its prospectus for a new public issue of shares amounting to a total value of 400,000 Marks. It was also accused of making false comparisons between the soil quality of the Meanja and Debundscha estates (*Graf's Finanzchronik*, 12 October 1908 to 10 May 1909). For a general overview of the tribulations of the Cameroon rubber companies see Rudin (1938), 265–269.

23. Heinrich Picht, by his own account, had, as a very young man, been briefly employed in Cameroon and had undertaken an expedition to recruit labour in the hinterland. After two years' local experience he had earned enough in commissions as a recruiter to come home and study law at Berlin and Paris Universities, after which he was invited back to Cameroon. He not only served on the Meanja Plantation Board and that of the Debundscha Plantation, but also by 1907 had taken over the direction of the *Deutsche Kautschuk AG*, in later years renamed the

Ekona Plantation (BAB, R1001/RKA 3539 and 3544). During and after the First World War he worked in the diplomatic service, took part in peace negotiations with the Allies, and in a trade delegation to Argentina. As Kemner also relates, he was one of the three agents (the others being Kemner and Kurt Woermann) sent to London by the combined plantation interests and the German Government to repurchase the plantations at the 1924 auction. He was also a director of the *Reichsverband der Kolonialdeutschen* (Wenzel, 1929). After 1929 he retired from the plantations to study medicine at Berlin, Innsbruck and Munich Universities, and in 1934 presented a thesis on the hygienic problems of large plantations in the humid tropics. The above information is partly drawn from the career details that introduce his thesis.

24. Renamed the Stauffenbergstrasse after the Second World War.

25. This information was kindly confirmed by the *Ettlin Gesellschaft für Spinnerei und Weberei AG*. The dates of his membership of the board are uncertain since its records were destroyed in the Second World War. See also *Deutscher Wirtschaftsführer*, ed. Georg Wenzel, 1929, and Julius Mossner, 1935. Esser was, in 1935, a member of the board of the Wintershall AG Kassel – no further information as company records were likewise lost in the war (Helmut Ernst, Wintershall AG, personal communication).

26. See Annual Reports of the WAPV (BAB, R1001/RKA 3504) which also refer to the progress of the replanted Ekona rubber plantation. See also Kemner, 1937: 182–83 for Picht's investments, and Epale, 1985, 86 ff., for post-1924 consolidations and changes in ownership.

27. The date of Esser's death was kindly supplied by the *Amt für Öffentliche Ordnung*, Baden-Baden and burial details from the *Stadtarchiv* Baden-Baden.

28. Max Esser's collection of *ethnographica* made during his travels was deposited in the Linden Museum, Stuttgart. It lacks the largest of the sculptures he collected, apparently stolen on his return journey. An interesting correspondence remains in the Museum's records, see Appendix II.

29. For a convenient contemporary exposition of the arguments, see P. Konings (1993), Introduction, 1–16, and the references therein. For the disruptive demographic effects of the coastal plantations on the local population, the Bakweri, see E.W. Ardener (1962), *Divorce and Fertility: An African Study*.

Mme. Dutreux, née Deichmann, Max Esser's mother-in-law with her baby son, born 1873

Lily Esser, née Dutreux, Max Esser's wife

Max Esser's father-in-law, 'Tony' Dutreux, photographed on a visit to Baden-Baden in 1927

The Château de la Celle St. Cloud, property of the Dutreux family

Figure 1.1: The Luxemburg connection: from the photograph album of Max Esser's daughter Elisabeth, made available by courtesy of Herr Josef Pfaffenlehner, Schloss Sandersdorf.

Figure 1.2: A Christmas Party: the Essers and the Ahns
(By kind permission of the Prinzessin Ratibor)

Figure 1.3: Max Esser and his daughter Elisabeth
(By kind permission of the Prinzessin Ratibor)

Figure 1.4: Robert Esser's 80th birthday in 1913: a family gathering in Baden-Baden (By kind permission of the Prinzessin Ratibor)

Figure 1.5: Number 6 Ahlsenstrasse, Berlin Mitte, Esser's residence between 1906 and 1917. The tiny figures on the upper balcony are those of his daughter and her governess, Mademoiselle Mion.
(From the photograph album of Max Esser's daughter)

Figure 1.6: The villa in Baden-Baden.
(By kind permission of the Gräfin zu Eulenburg)

Max Esser with Prince Adolf von Schaumburg-Lippe on the race-course at Norderney, 1913

Max Esser in uniform at Guben in 1915

With Freiherr von Solemacher in Brussels in 1916

Figure 1.7: From the middle years of Max Esser (All from his daughter's photograph album, made available by courtesy of Herr Josef Pfaffenlehner, Schloss Sandersdorf.)

A Key to Plates 1.2 and 1.4

1.2 A Christmas Party: the Essers and the Ahns.

The encircling group of adults, L. to R.:

1. Max Esser; 2. Karoline von Caro, his sister, wife to Georg von Caro; 3. Ferdinand Esser, his elder brother; 4. Henriette ('Harry') Ahn, his sister, wife to Albert Ahn; 5. Robert Esser senior, his father; 6. Adele Esser, née von Kaufmann, his mother; 7. Georg von Caro; 8. Maria Esser, wife to Ferdinand Esser, née Kreuser; 9. Albert Ahn (hand in pocket).

Max Esser's nephew and nieces, L. to R.:

10. Robert Esser junior, son of Ferdinand Esser; 11. Gertrud Ahn, later Merrill, daughter of Albert Ahn; 12. Carola Ahn, later Freifrau von der Schulenberg, daughter of Albert Ahn; 13. Else Esser, later Massen, daughter of Ferdinand Esser; 14. Adele Ahn, later Meynen, daughter of Albert Ahn.

The children with linked arms in front:

15. Adele, daughter of Karoline by her first husband, adopted by Georg von Caro; 16. Wolfram Ahn, son of Albert Ahn; 17. Elisabeth Esser, daughter of Max Esser.

1.4 Robert Esser's 80th birthday, 1913 – a family gathering.

L. to R. from top to bottom:

Top Row: 1. Henriette Ahn; 2. Schumacher (*connection unknown*); 3. Max Esser; 4. Robert Esser junior.
Second Row: 5. Adele Ahn; 6. Karoline von Caro; 7. Else Esser; 8. Gertrud Ahn; 9. Albert Ahn; 10. Ferdinand Esser.
Third Row: 11. Adele Esser; 12. Robert Esser senior; 13. Carola Ahn.
Bottom Row: 14. Elisabeth Esser (holding stick); 15. Adele von Caro.

Part II

Esser's Travels

2

The Outward Voyage

[Max Esser begins by explaining the task he had set himself. The first period of doughty inland expeditions and scientific discoveries had ended. Now it was time to start on practical and profitable enterprises, and to study the prospects for successful investment in the colonies. Such a study called for examination of the economic development of the long-established Portuguese colonies: the islands of Sao Thomé and Principe, and the Province of Angola offered models of use to Cameroon and German South-West Africa respectively. When his proposal for a fact-finding expedition became known Victor Hoesch and Dr Eugen Zintgraff were ready to join him. Dreams of Solomon's Gold had ceased to haunt the European imagination, which was now engaged by Africa's astonishing fertility. Germany was now in a position to exploit its tropical colonies but more information was needed if lives, skills, enterprise and capital were not to be wasted.]

Of late, Africa has astonished the world by its unimagined fertility. African cocoa, sugar, coffee, tobacco, cotton and other crops are beginning to gain the first place on the World market and offer serious competition to the produce of the Americas and Indies... As to the fertility of the soil, yielding up to three harvests a year, Germany has come off very well in the greater part of its colonies. I am convinced, on the basis of my thoroughly conducted studies made on the spot, that the time will not be long before dazzling results are available to prove that German blood, German energy and labour, German enterprise and capital have not been sacrificed in vain on African soil.

[With these thoughts in mind, Esser boarded the Portuguese vessel *Cazengo* in the Spring of 1896 with his two companions – and his fox-terrier Scherz, who was taken on board in the face of much opposition from Portuguese officials.
 On the third day out to sea and a visit to the picturesque port of Funchal, they anchored in Madeira, from whence they sailed south-west to the Cape Verde Islands, which they reached on the 12th of May. Here they made some excursions and were advised by the famous Portuguese explorer Serpa Pinto. Esser does not fail to list their particular exports. After a gruesome incident here, Esser, a shocked onlooker, ponders on the character of the Africans.]

My free time on board I used as much as possible to study the character of the Negro. I will never forget one incident at the Cape Verde Islands. A canoe, rowed by three Negroes and carrying a cask of oil, came to our ship. When the men tried to lift up the cask to the ship's gangway, the boat capsized, and the Blacks together with its paddles and the cask fell into the water. In the immediate neighbourhood there were at least ten manned boats. Their crews themselves recovered the paddles and the oil cask, but did not bother at all about the Negroes struggling with the waves. Two of them managed to reach the ship's ladder, the third, however, suddenly drowned with a scream of terror. A shark had seized him, the water turned red, and then a whole shoal of them arrived and soon finished the man off. This was a terrible sight that was clearly observable due to the perfect transparency of the water.

After the event, I asked one of the Angolans employed on board why nobody had tried to help the poor wretch. The man was quite surprised and remarked that the cask of oil would be of more value than the rascal who might have well caused the boat to capsize. His fate was a due punishment!

> [Esser's dealings with those Africans who had been in close touch with European commerce had not been happy so far, and it seemed to him that they lacked a sense of duty and neighbourly love. After leaving Sierra Leone behind them, they turned east into the Gulf of Guinea, and sailed past the ruins of the old Brandenburg fort of Great Fredericksburg, founded in 1683.]

After difficult and lengthy negotiations with the Dutch, on the New Year's day of 1683, Major v.d. Groeben hoisted the flag of Brandenburg at the spot where soon after the fort of Great Fredericksburg was constructed. Two hundred years later the German corvette *Sophie* visited the ruins of the fort and brought one of the old Brandenburger canons back to Germany as a memento . . .

In 1684, a deputation of Negroes from the Brandenburg colony had visited Berlin to pay homage to the Great Elector Frederick. In the meantime, the protectorate grew to a considerable extent . . . In 1687, Brandenburg had acquired another colonial possession, the Arguin island situated south of Cape Blanco, and traded successfully in rubber.

Nevertheless, and particularly because of Dutch intrigues, Brandenburg's first colonial enterprises had not been full of blessings. Their decline commenced after the Great Elector's death . . . Both possessions were sold off to the Dutch West Indies Company in 1720 for 7,200 ducats and 12 slaves, six of whom were in golden chains.

The Dutch company did not hold on to these possessions successfully either. At Great Fredericksburg, Jan Cung, a loyal Negro prince, led a fierce struggle for years against the Dutch in order to stay under Prussia. It was only in 1725 that the Dutch were able to take possession of the fort. The promising colony had always been neglected by the Dutch until it was sold to the English in 1771, under whose rule it started to prosper. The Dutch administered Arguin even worse than Great Fredericksburg... In 1721, it finally fell to the French[1] ...

Thus, the colonial enterprises of the Great Elector, that had started so promisingly, had now collapsed into nothing. Germany's position in the world as a colonial power would have been different had Frederick's successors maintained his colonial policy!

Notes

1. Esser relies for his remarks on Great Fredericksburg on Koschitzky (1887), Part I, 55–66. Max von Koschitzky was engaged early in the field of revivalist colonial history.

Figure 2.1: Part of the ruins of Great Fredericksburg
(from Wilhelm Kernner's photograph album)

3

Sao Thomé and Principe

[Two days later they came in sight of the first islands of the Gulf of Guinea, uninhabited, and all extinct volcanoes when first discovered by the Portuguese navigators in the 1470s. The Portuguese called them the Line Islands. They were initially settled with convicts and slaves. Sao Thomé, as we shall see later, had a chequered past.

Esser supplies the reader with a brief geographical introduction which stresses their common volcanic origin with that of Mt Cameroon, their cover of tropical forest and the climatic features which made them so suitable for plantation agriculture.]

Taking a long view, one Dr Matheus Augusto Ribeiro, a modest country doctor, had identified the productive capacity and the enormous progress that the plantations on Sao Thomé would undergo once public order had been established. He arrived in the early 1870s and spent day and night on his mule visiting his extensive plantations. Tirelessly, he thought of nothing other than planting and planting, as much as he was capable of. He put every penny of his savings into improved plantings. Even his wife and children took part in planting cocoa. In 1888, after 16 years of unremitting work he was able to sell his interests to a front man of the *Banco Ultramarino* for some 6 million Marks. Today the value of Ribeiro's properties has risen to many more millions. Due to Ribeiro's constant admonitions, instructions and his personal example, on Sao Thomé today we find a flourishing plantation enterprise, initiated with outstanding success. It makes the whole island look like one big plantation.

The plantation industry employs some 50,000 people, including 5,000 Whites and 1,000 Chinese. The rest are either Krus from Liberia, Angolans, or come from Sierra Leone and Cabinda. The labour system installed after the alleged abolition of slavery promised the African labourers, most of whom signed a five-year contract with the plantation manager, free passages to and from home, free food, and 100 Marks per annum in wages, which was less often paid in cash than in cloth and spirits.[1] Free food consists of 700 grams of rice a day, local fruit and vegetables, and a pound of meat a week. Work begins at 6 a.m. and lasts until 11 a.m., when there is a two-hour break, and continues from 1 p.m. to 6 p.m. Sunday morning is worked, the afternoon is free. No holy or saint's days are observed despite the prevailing Catholicism.

The main crop cultures are cocoa and coffee which grow excellently. The following figures will illustrate how the principal exports of the island, particularly cocoa, have developed in the last twenty-five years. In 1869, 2,081,712 kg of coffee and 50,867 kg of cocoa were exported. In 1896, the coffee exports amounted to 2,960 654, and the exports of cocoa rose enormously to 5,670,000 kg. The total value of exports from the plantations rose from 1,600,000 Marks to 15,350,000 Marks over the same period. As cultivation increased, trade grew proportionately year after year. The value of imports rose concomitantly, from 660,000 Marks in 1869 to 5,050,000 Marks in 1896.

Neither Principe nor Fernando Po had experienced such a dramatic revival as Sao Thomé, but efforts were now being made, with greater zeal than hitherto. For example, in 1895, the *Companhia da Ilha do Principe* planted 330,000 cocoa and 30,000 coffee plants, and 50–60,000 sugar canes on this island, and planned further cocoa plantings in 1896 . . .

I should add that I was able to convince myself personally of this high output. These figures are easily ascertainable from the lists prepared for tax purposes, which are widely known. There are no land or property taxes levied, but only an export tax collected at the port. Therefore nobody has any reason to hide such figures . . .

Among other entrepreneurs is J.H.M. Mendonca, who owns the 900-hectare plantation of Boa Entrada. A hectare of virgin forest is acquired at present for 200 to 220 Marks. In 1894, he made a net profit of 350,000 Marks from his production of 460,000 kg of cocoa and 40,000 kg of coffee . . . In 1876 Monte Café, another property, was bought by the brothers Chamisso for 350,000 Marks. The present owners were recently offered 5 million Marks for the property, by no means too high a price given its good results. Average profits over the last ten years have exceeded 500,000 Marks and the 1896 harvest of coffee, cocoa and cinchona-bark . . . has brought in gross profits of 800,000 Marks. This was only the beginning of its development: the 10,000-hectare tract was not yet fully planted up.

As a curiosity, I will mention the rise of the Conte de Valle Flor who is today one of the biggest landowners of the island. He has an annual income of $2\frac{1}{2}$ million Marks and cuts a grand figure in Lisbon. As plain 'Mr John' he ran a small bar and had a business in dried fish on the island. He had put his modest profits into cocoa, planting a hundred bushes a year, then a thousand, until he became a big landlord and a Portuguese Count. None of this involved speculation but was owed merely to good use of the land.

One must see the fine plantations of Sao Thomé and Principe for oneself and walk day-long through them to gain a true impression of the wealth they represent. How depressing it was . . . to find the German colonies constantly attacked in parts of the German press and in the *Reichstag* as the worst of the lot . . . and by contrast how pleasant it was to be able to proclaim their beauty and wealth! . . .

The *Cazengo* anchored in the beautiful bay of Principe, surrounded by thickly forested cliffs . . . It was six o'clock in the evening, and getting dark . . . From the forest we heard the shouting of the monkeys and the parrots. The moon provided a broad silver-bridge, inviting a visit to the island . . .

The next day was marvellous too. The ship lay quietly in the bay. Giant sharks circled around in search for prey . . . Soon the Negroes came in their typical dugout canoes to our ship, and soon a trade in fruit developed.

When we went ashore, we passed the old fort, halfway up a mountain at the entrance to the bay. From the ship we had already seen the Portuguese flag on its summit. We were surprised to discover that it lay in ruins, overgrown with palms . . . Our boatmen showed us the old cannon barrels which had tumbled down the slope, lying on the sea-bed. I counted seven of them . . . Later I noticed that a number of others, all good bronze, were used on the island as boundary markers . . . The garrison of 25 men seemed happy to have got rid of them: their leader confirmed that he sees their sole task as local policing. Everything seems to be in ruins; even the four churches . . . their floors and roofs torn apart by the strong upward thrust of palms. Even in the pulpit there were palm trees . . . We asked the inhabitants why the churches were abandoned. We were told that they had neither the time nor the money for their restoration, and that the missionaries had long ago died of fever. The regime cared as little about the churches as they did about the cannons.

Nevertheless, the Island is not God's stepchild despite the neglect of His servants and its Government. It is richly endowed by Nature. The notion that the forests contain poisonous plants that, in the long run, can endanger the indigenes' lives is hardly credible. As the day went on, we saw a number of well laid-out coffee and cocoa plantations as well as banana groves. Pineapples and avocados abounded. We gathered them in the woods, and washed them down with milk from the coconuts picked for us by the Blacks – a trouble-free picnic . . .

Next day some wealthy, well-dressed Negro families came aboard for the trip to Sao Thomé. The ladies' toilettes would have caused a stir in Berlin or Paris. The smartly got-up ladies, in fashionable puffed sleeves and iridescent gowns in loud colours, which did not suit their square, heavy-set figures, gave me a lot of quiet amusement. Their children, in sailor suits, made a quite different, indeed a charming impression. They greeted every European with a 'good morning' and kissed their hands, showing how thorough their training in good manners had been. The town of Principe was not noteworthy, consisting mainly of grubby hovels and a single large building accommodating its government, guards and the post and telegraph office . . .

The anchorage of Sao Thomé, though not so picturesque as that of Principe, was wider and deeper, and to the north of it lay a string of islets – the aptly named Goat Islands – on one of which stood the lighthouse of Sao Thomé. During our stay we were the fortunate guests of the German Consul, Herr Spengler, who had come to meet us. Even so, we had to spend four hours going through Customs before our baggage was cleared.

Spengler is regarded as one of the leading German businessmen in West Africa: as the manager of the Monte Café plantation, he is widely known outside the island for his thoroughly professional knowledge of tropical agriculture. His plantations are generally regarded as constituting a first-class model to be copied.[2] After a slow two-hour ride on his Spanish horses, followed by the carriers with our loads, we reached his ideally placed residence, 700 metres above sea-level . . . The marvellous view from the residence led down to the sea, across the fruit-drying yard facing the house and extended over endless plantations, banana groves, orange trees, virgin forest . . .

Behind the house is an extensive botanical garden with a huge variety of plants . . . Rare fruit trees grow beside numerous vegetables . . . varieties of tea, coffee and cocoa, spices, nuts, fibre-plants, oil-bearing trees, rubber vines, grain crops, legumes, even incense plants, had been collected, and were laid out in beds garlanded with roses . . . Among these specimens are the Emperor palm which provides the raw material for Panama hats, and the avocado pear, relished by man and horses, donkeys, oxen or dogs alike ...

Sao Thomé had been a sugar-producing island till the end of the seventeenth century. With the Portuguese settlement of Brazil ... the settlers abandoned it to try their luck over there. Principe remained in use as a trans-shipment port for African slaves shipped to the Americas, and to begin with, for the few slaves sent to Europe. Sao Thomé was wholly abandoned. The islands remained neglected, until, in the 1820s, a start was made with coffee planting, overtaken in the 1860s by cocoa. A little sugar is still grown and mainly used to make rum in the Islands themselves for local consumption.

One of the other local industries was the manufacture of dugouts from the huge silk-cotton trees that abounded on the islands. I observed hundreds of natives engaged in bringing one down to the coast on an ingenious slide made of smooth palm stems acting as rollers and guided by tow-ropes of twisted lianas.

On the afternoon of Whitsun, we made an excursion to outlying plantations, planning to use the 56 cm. gauge light railway ... Despite its being a church holiday, the islanders were at work till midday. Then there was a general inspection: the labourers were assembled and the cleanliness of their clothes and feet was examined. Those who passed muster were rewarded with a leaf of tobacco and a glass of wine.

After the inspection, Herr Spengler, his brother-in-law, Herr Seixas, Zintgraff, Hoesch and I, mounted the light railway, with Herr Seixas in charge ... On a slope, the speed of the wagons increased; it was terrifying ... the brakes were of no avail. We soon discovered the reason: the track had not been used for some days, the sap of overgrowing plants had made the rails slippery. The wagons derailed at a sharp corner and covered us with their remnants ... Hoesch was the most seriously hurt ... with some deep gashes on his right arm, his legs and his head. Soon, after our wounds were bandaged, we returned home, and were able to joke about our equatorial derailment ... and our lucky escape.

During the next few days ... Spengler conducted us – Zintgraff and me – around to study the plantations ... this time on horseback. With increasing admiration, we inspected them, including one devoted to cinchona. We learned too, in the course of our study-tours, of the history of Sao Thomé, which was first used to house white convicts ... each of whom was provided with a black slave wife. Slaves taken from the mainland were then settled ... All these settlers found subsistence relatively easy. The Government and clergy attempted to establish some sort of order among the inhabitants, but their efforts were of little avail. The mixed race that had emerged seemed to combine the faults rather than any of the merits of the two peoples. They gave themselves over to every kind of violence . . . and rebelled against the Governor. In the middle of the sixteenth century a new element was added to Sao Thomé in the form of 200 shipwrecked Angolans, against whose banditry the Government seemed powerless . . . This went on for a hundred years and more, and the disorder was such that the Portuguese were happy enough when, in 1778, the Spanish Crown relieved them of the responsibility for Fernando Po and Annobon.

The circumstance that Governors of the islands were from time to time excommunicated by the Catholic clergy, while the Governors in turn exiled members of the clergy from one island to another, perhaps illustrates the desultoriness of their governance. This state of affairs ended in the last quarter of this [the 19th] century[3] . . .

In Fernando Po, the situation had also improved. In 1876 it was offered to Germany for 300,000 Marks through agents, at a time when the Spanish Government had no idea of the fertility of its soils. Unfortunately, the same was true in Germany then, as it is now among those opponents of colonisation who still regard the Cameroon coast as a valueless asset.

Notes

1. For a study of hidden labour costs in the Portuguese islands see W.G. Clarence-Smith in *Portuguese Studies,* 6, 1991, 152–172.

2. Spengler had already made an exploratory visit to the Cameroon Mountain area and had reported favourably on its suitability for the cultivation of cocoa and coffee: see *DKB* 5, 1894, 282–288, and Edwin Ardener, *Kingdom on Mount Cameroon,* 1996, this series, p. 106.

3. Compare Sir Harry Johnston (for whose career see R. Oliver, 1957) who visited the Portuguese African possessions in 1882–3, and gave lectures on them to the Society of Arts on 12th February, 1884 and to the British Association in 1885. The latter was published in the magazine of the Scottish Geographical Society of October, 1885. The former, which contains a vivid and sympathetic description of Principe and Sao Thomé, is quoted at length in *Portuguese Planters and British Humanitarians: the case for S. Thomé,* 1911, 68–74, (trans. I. A. Wyllie). This compilation was published by the Lisbon newspaper *Reforma* in rebuttal of proposals by

W. A. Cadbury, H. W. Nevinson and others, including German chocolate manufacturers, to boycott Sao Thomé 'slave-grown' cocoa. It includes a letter of February 1910 from the Chairman of the WAPV and Esser's successor, Kemner, recording their favourable impression of labour conditions on the two islands.

4

Cameroon – the Historical Background

[In this next chapter Esser takes his readers helter-skelter through the history of the African coast, starting with the legendary tale of Herodotus about the Egyptian circumnavigation of Africa, and the mythical voyage of Hanno, to Dapper's account of the discoveries of the West African Coast. This is evidently a scissors-and-paste job, derived from secondary sources, to judge from the misprinting of Dapper's name. We reach rather firmer ground with the treaty made by the British in 1841 with the Duala King, Bell, and then Akwa (in 1842) to abolish the slave trade (a somewhat oversimplified rendering) and pass rapidly on to the earliest German trade settlement via a brief mention of Duala internecine conflicts and 'an English mission', the London and Jamaican Baptists. The first German settlement was that of the Hamburg trading firm of Woermann in 1868, followed in 1875 by Jantzen and Thormählen, who had been in their service. We are given Thormählen's own retrospective account.]

Recently, Mr. Thormählen has commented on his first voyage to Cameroon: 'In the autumn of 1864, the Woermann House, which had traded successfully in Gaboon . . . decided to start trading in the Oil Rivers. A suitable ship was acquired and a cargo destined for the West African trade put together. I set off in early 1868, on the *Titania*, flying the black-white-and-red Hamburg flag. Arriving in spring at Fernando Po, I decided, on the basis of local information, to make for the Cameroon estuary rather than for Old Calabar. There I found six or seven English vessels . . . already at anchor; the Cameroon estuary had long been frequented by Liverpool and Bristol traders who controlled the whole trade. They conducted this trade, not from permanently moored hulks, such as were already used in the other Oil Rivers, but from so-called transit ships which, once laden, could return home. Masts and cordage were taken down and a matting awning covered the decks . . . while goods were traded. I soon decided that a permanent hulk would offer more advantages. At the beginning, the English did not seem to care about my activities ... When they realized that I was a competitor, they started to plot against me ... Indeed, when the British Consul had, in 1872, closed the river to trade after a fight between an aggressive English trader and the incensed Joss people [one of the Duala groups], I insisted on my rights to pursue my legitimate business in palm oil and ivory ... and that the decisions of the Court of Equity were not binding on me. The Consul declared that he had the power to prevent my trade. I had to admit this, but added that if he would try to do so, it certainly would be a good business for me. So he did not go beyond intimidation ... The notion that Germany had become a power to be reckoned with was now well understood ...'

So much for Thormählen. By 1874 the Woermann firm had established its West African steamship line, and had reported the need for a German consul in Cameroon. The same memorandum also underlined the problems of trade with the hinterland, which was controlled by the Duala ...

In 1882 a few of the Duala chiefs had invited the establishment of an English protectorate. As they obtained no answer they turned to Germany and, with the help of Woermann, as representative of the German firms in West Africa, the Imperial Government despatched Dr Nachtigal with powers to proclaim a German Protectorate ...

It was not difficult to get the agreement of the coastal chiefs despite the intrigues of the English traders, highly interesting to observe. Dr Nachtigal arrived on the *Möwe* on 12 July 1884, and two days later the German flag was ceremonially hoisted first in Bell Town, then Akwa, and then Deido Town. English protests followed ... and local intrigues continued, the proof of which I can illustrate, for example, by means of a letter from the British Consul, Hewett, to the Chief of Bimbia[1] ...

[The English determination to maintain their trading position led to perpetual difficulties being made for Germany. Esser cites the support supposedly given by the English to internecine Duala quarrels and to the expedition of the Pole, Szolc-Rogozinski ('who apparently originally bore the noble name of Scholtz') to Mt Cameroon.

In October 1884 the Imperial Government announced the annexation of Cameroon to foreign powers through its envoys, but certain English interests had not given up hope that they could bring about the abandonment of the colony by Germany. Local affairs had taken a bad turn with some direct conflicts between natives and Europeans, menacing to the latter, as well as a heightening of bitter strife between the Duala dynastic groups themselves.]

Luckily at the end of December a part of the West African squadron under Admiral Knorr arrived ... The hostile chief Akwa had torn down the German flag and threatened German factories ... A landing was made with a detachment of four machine guns ... with orders to attack the villages of the disobedient chiefs and capture them if possible ... but they escaped. The British Consul Hewett made vain protests, but England was in no position to arrest the development of the German colony ... To elaborate on the acquisition history of the various parts of the hinterland would take me too far here. I will just add that where the Germans dealt with the French, south of the colony, all went easily, but to the north of it every inch was acquired with difficulty to the extent that in Ambas Bay Admiral Knorr was ordered to take Rogozinski prisoner. The energetic action of the authorities eventually resulted ... in the cession of the English Baptist properties in Victoria to a Swiss Mission[2] ...

By May 1885, the British had recognised the rights of Germany in Cameroon, with a boundary at the Rio del Rey, while the Rio Campo formed the boundary agreed with the French. A mixed commission was currently undertaking further detailed boundary demarcation ...

The first German Governor was Freiherr Julius von Soden ... At his side was the junior barrister Jesco von Puttkamer, the present Governor ... Von Soden soon realized how suitable the territory was for plantations[3] and welcomed the first enterprise, the Hamburg *Land- und Plantagengesellschaft*, set up in Man o' War Bay. During von Soden's time the first steps towards orderly government were made – an advisory council, under the Governor, an arbitration court, and the beginnings of a customs administration were set up, and the trade in spirits and arms was brought under control ...

To be sure risings had taken place as everyone knows, in 1886, 1887 and at Christmas 1893, principally caused by the attempts of the Government to remove the monopoly of the coastal tribes over relations with the hinterland tribes. Granted that these battles have ended with some disturbing losses, I am of the opinion that the time of disturbances is now over and peaceful development can take place.[4] The administration of justice was now orderly, at least in the directly administered areas and in those under District Officers. A regular administration in the hinterland was still a long way away, but the establishment of so-called Native and Arbitration courts in 1894 and 1895, albeit with limited powers, under particular chiefs, has been established in the administered areas, with a chairman and a court clerk. The members of the arbitration courts act as advisers to the administrators acting as judges in criminal cases. Despite the present limited authority of these courts they are of the highest importance in establishing and developing the notion of the rule of law among the natives, and will have a positive effect on the organisation and conditions of labour . . .

Finally allow me a few words about developments in the regulation of land matters because of its great importance for all those who plan to engage in plantations . . . A decree of 2 July 1888 established Prussian law as the basis of land acquisition and other connected material obligations within the range of consular jurisdiction. However, this had no application to native lands in which, as a rule, established custom and usage was expected to prevail. For the White inhabitants a land register had already been established in which all non-native land acquisitions had to be entered. To the Governor belonged the right to lay down the conditions under which unoccupied land or land bought from natives by Whites could be acquired. It followed from this that all agreements of this kind required the Governor's approval. This was to protect the natives from disadvantageous deals. If the purposes for which the land had been registered – for example, for plantations – were not put in hand promptly, the land so acquired became Government property after a specified period . . . In order to protect the natives still better from deals which might disadvantage them, it was further laid down, in December 1894, that a representative of the Governor must be present at their conclusion; likewise any appropriation of unoccupied land had to be notified to the Government within six months, failing which all title to it was lost. In short, title to land implied its prompt productive use for plantations.

In the summer of 1896, a new decree was issued concerning the creation of Crown Land and the acquisition of parcels of land. This claimed all unoccupied land as Crown Land, and 'unoccupied' land was defined as land not in the possession of the Government, private persons or natives . . . The Crown Lands, where Land Registers already existed, were to be recorded and might be alienated by the Governor.[5]

It can be seen that these enactments protect the rights of both natives and colonizers. With these regulations, a favourable basis has been created for the economic development of the protectorate. Since secure title is essential to plantation enterprise, the laws and regulations of Cameroon exemplify this security in the highest degree.

Notes

1. Here again Esser seems to rely on Koschitzky and refers to him in dealing with Consul Hewett's correspondence with Bimbia (1887, 142f.).
2. For a well referenced study of Bimbia 1833–1879 from British sources see Lovett Z. Elango in M. Njeuma (ed.) *Introduction to the History of Cameroon*, 1989, 32–62.
3. For a convenient coastal chronology see E. Ardener, 1996, Appendix B, 351 ff., and for a re-examination of the historical evidence for the coast between Rio del Rey and Duala, A.D. 1500–1650, see the same volume, 1–40. Thormählen had represented the C. Woermann firm between 1868 and 1874 and had subsequently been kept informed by his own representative, Johannes Voss: see *Mitt. Geogr. Gesellschaft in Hamburg*, 1884, 328–334.
4. For Cameroon history from its discovery to 1886 Esser has evidently made use of Koschitzky (1887) Part 2, 124–180. For a collection of contemporary accounts, see S.G. Ardener (1968), *Eyewitnesses to the Annexation*, reprinted 1996 for the National Archives Buea. Also Hugo Zöller (1885), Max Buchner (1887) and *Aus dem Tagebuch von Eduard Woermann*, in Hans Zache (ed.) *Das deutsche Kolonialbuch*, 1926, 261–263. The problems faced by the Basel Mission in taking over the London Baptist Mission properties in Victoria (Limbe) are described in W. Schlatter (1916), Vol. 3, 213–242. For a view of the trade, politics and internal rivalries of Duala, see A. Wirz (1972), 36–91. H-P. Jaeck, in H. Stoecker (ed.), *Kamerun unter deutscher Kolonialherrschaft*, Vol. 1 (1960), 29–95, who provides a fully-documented account from German archival sources including Woermann's memoranda and lettters; for the wider diplomatic setting see H. Ashby Turner in Gifford and Louis (eds.), *Britain and Germany in Africa* (1967), 47–82. British sources are dealt with by V. G. Fanso in his contribution to M. Njeuma (1989), 63–87. For documentary references see Latham (1973), 89–90 and 139–142, and Austen and Derrick (1999).
5. Indeed von Soden had speculatively acquired a tract of land in the Buea area in 1887 which he ceded to the WAPV, of which he became a director ten years later. See K. Hausen (1970), 306, 312–314, Clarence-Smith (1993), 191, who quotes ANY, FA 1/317 and 1/350, and Ardener (1996), 117 ff., who describes the gerrymandering of boundaries to von Soden's benefit by Dr. Preuss on the instructions of Governor von Puttkamer.
6. This was an optimistic assessment, as a reading of the DKB for the next few years will indicate. Some of the Bakweri of Buea remained recalcitrant and had to be moved in 1899. Early in the same year a mutiny by Vai soldiers in Buea was narrowly averted. Punitive expeditions fol-

lowed the violent deaths of Lt. Queiss and the trader-recruiter G. Conrau. For further references see J. Ballhaus in Stoecker (1968), Vol. 2, 130–162 and Chilver (1967b) in *J. Hist. Soc. Nigeria* 4(1), 155–160. For retrospective accounts of the Cross River and adjacent groups which rebelled between 1899 and 1905 see Lessner (1904) in *Globus*, 86, 273–278, 337–344, 392–397, and A. Mansfeld (1908), 17–21. For contemporary official accounts of the later Cross River risings see *DKB* 16 (1904), 185, 189–192, 481, 598 ff., 701–702, 735.

7. For Governor von Puttkamer's attitude to 'land in native possession' and 'unoccupied land' see his *Gouverneursjahre* (1912), 103–104, in which he scornfully dismisses long-fallow systems of peripatetic forest clearance and female cultivation as primitive and wasteful. See also, in relation to the Bibundi plantation lands, the exchange between Seitz and Esser in BAB, RKA 3513, 29, in which Seitz defends scattered small settlements against Esser's proposals for their consolidation into larger villages.

On the imperial decree of 1896 and its application see Rudin (1938), 367–378 and 398–399, Clarence-Smith (1993), 197–199, Ardener (1996), 151–155, and Eckhart Rohde (1997), 69–110, the last with full references to German archives on land policy in the ANY.

5

Land and People in Cameroon

[Esser next introduces his readers to the geography of 'old' Cameroon to the extent it was known at the time. The Cameroon mountains and the Atlantic coastline are briefly described, but what was of more significance was that the explorations to date had made it clear that beyond the coastal plain lay two successive terraces.]

From the lowlands south of the Cameroon river, the inner plateau rises in two terraces that reach the height of 700 metres rising precipitously in gigantic steps. Unfortunately they bar the navigation of Cameroon's rivers on their middle and upper courses, since, when the rivers pass through the terraces, they form waterfalls and rapids. Often these can be bypassed without major difficulties, although they appear as tiresome obstacles. At the northern boundary of the protectorate, partly forming the boundary itself, runs the Rio del Rey. To the south of it the Mungo, the Wuri with its tributary Abo, as well as the smaller rivers Lungasi and Denga discharge themselves into the Cameroon bay. In its lower course, the Mungo is navigable for small steam boats. There are however rapids on its upper course from Mundame upwards...

The coastal plain and the approaches to the inland plateaux are almost completely covered by thick forest. The wilderness can only be penetrated by river, by so-called elephant tracks or by the narrow paths of the natives. This is so because the gigantic trees are tied together by countless lianas running from trunk to trunk while thick undergrowth prevents any progress off the track. The forest is followed by a transitional parkland zone marked here and there by very fine gallery-forests. On the plateau savanna itself, palms are still frequent... including the oil palms which are of the greatest importance to the natives... The wildlife increases in number and variety as one penetrates inland, with the larger antelopes, buffaloes and elephants to be seen... though the last are now rarer in the north than in the south, where they are still frequent enough to sustain the ivory trade for some time to come. Apes of all kinds . . . a rich array of marshland avifauna . . . and, in the northern hinterland, predatory animals, of the commoner African kinds, abound.

> [Esser now introduces us to a contemporary racial and ethnological perspective, pretty typical of popular German views of the period.[1]]

The Cameroon population is divided between two great and distinct families . . . sharply divided by speech, custom and appearance. The Bantu presents a typical Negro appearance with protruding buttocks, thick, bulging lips, flat feet, and long ape-like arms. The Sudanic Negroes seem a much handsomer people: many of them are Moslems while the Bantu are pagans. The Sudanics are prouder than the Bantu and therefore less willing to work for others. It looks as if the Sudanics, who only reach up to the central hinterland, are falling in numbers owing to the unnatural disposition of women, who are to be found in large numbers in the royal harems.

The coastal people are Bantu. Some of them, such as the Duala, beyond them the Mungo and Wuri folk, and further south the Balimba, Bakoko and Batanga, are usually traders, while those to the north of them are engaged in agriculture, stock-keeping and hunting. The intellectual aptitude of the Bantu is by no means lacking. They learn fast and do so with great enthusiasm.

The Duala, who formerly had the middleman trade with the inland tribes wholly under their control, are generally lazy but also crafty and treacherous so that Europeans must be on their guard if they are not to be defrauded. Their monopoly of the inland trade has happily come to an end as a consequence of the German expeditions into the hinterland. They are, so far, still disinclined to take on regular work and the Government must therefore regard accustoming them to it as one of their first tasks and duties.[2]

To the extent that the natives are cultivators, they are subsistence farmers growing chiefly bananas, cocoyams, yams, sweet potatoes, coconuts and sugar cane. The Sudanics grow maize, rice and sorghum. [Other kinds of millet, bulrush and finger millet, were grown in the hinterland]. Animal husbandry is at a low level although fine sheep, goats, and cattle are to be found. The Government is concerned with improving this state of affairs, and research is ongoing.

Industry hardly exists though weapons and cloth are manufactured in the hinterland. The coastal folk can obtain all they need easily and cheaply from the Europeans so there is obviously no prospect that a locally-based industry will develop.

The Europeans are still few in number, barely 200 including six women. Most are Germans, then follow English, Swedes and finally a few Swiss, Americans and Spaniards.[3] The greater part of the Germans, perhaps 150, are engaged in trade in the coastal trade counters – the 'factories'; about 30 are employed by the plantations.

[A rather perfunctory excursus on health and climate follows. Quinine and daily baths, he writes, are essential – but Esser soon returns to trade.]

The chief exports engaging the established traders in Kamerun [Duala] and Great and Small Batanga in the main are palm oil and palm kernels, wild rubber, ivory, ebony, piassava and the plantation-grown cocoa, coffee and tobacco, while imports consist of cotton goods, ironware and furniture, instruments and machines, haberdashery and linen, copper and brass goods, salt in large quantities, and finally spirits, weapons and gunpowder, the last three paying a heavy duty.

The main difficulties facing the development of trade into the hinterland is the absence of good routes and, more especially, the difficulties placed in its way by trading tribes. The waterways, as we have seen, are only partly usable and in the future it will be necessary to improve them. The construction of roads into the hinterland is another necessity and more feasible than generally assumed.

But however important trade is for the Protectorate its future depends on the plantation industry and it was to study this proposition, set out by the leading expert in this field, Professor F. Wohltmann in his recent book, that was the main aim of my visit.[4]

[Esser then takes us on an elementary trip round the differing requirements, in temperature and rainfall, of both European and tropical crops grown under plantation conditions.]

The area around the Cameroon mountain has a distinctive tropical climate ... The temperature, recorded between a highest point of 35°C and 15°C at its lowest, is generally well-tempered. The rainfall is equally favourable not so much because of its quantity but because of its seasonal distribution, with only two totally dry months. In short, despite a total annual rainfall of 3,000 mm ... there is no waterlogging or excessive damp ... Professor Wohltmann regards the area as offering the best hopes for cocoa, vanilla and coffee combining, as it does, a typical tropical climate with a moist, warm greenhouse atmosphere. ... Moreover, according to the Professor, the well-weathered volcanic soil, low in lime but with a high nitrogen content, was a special gift of nature widespread in this region, especially suited to tree and bush crops of a perennial character... The combination of climate, soil and the contiguity of the sea, requiring only a short route to cargo-loading facilities, is especially fortunate...

So it is odd, to say the least, that some circles in Germany believe that it has occupied an empty desert.

Among the suitable crops one can name cocoa, coffee, tobacco, sugar cane, vanilla, ginger and cinnamon, but cocoa must come first as there are few places on earth so suitable for its cultivation.[5]

The Government is well aware of the advantages of plantations and has set up several research stations... the most important of which is the Botanic Garden in Victoria, at the foot of Mt Cameroon... and on an inlet that provides a convenient anchorage, and which could be developed into a good harbour. It is only four hours away by ship from Kamerun [Duala], the seat of government... It has, in a short time, been remarkably developed by its director, Dr Preuss... 27 hectares of its 50 hectares area have already been planted. Unfortunately, it still lacks a laboratory, which he badly needs if he is to perform his function of advising planters. Twelve different kinds of cocoa are being tried... The *arabica* coffee is already doing better than the Liberian, and two local wild varieties... are being tried too. Various kinds of rubber-bearing plants are being tested, especially since exports have been reduced because of the destructive methods of collecting wild rubber favoured by the natives... Likewise, spices and local fruits, in many different varieties, are being tested... An animal husbandry research station... has been established under Herr Leuschner at Buea, as well as a convalescent home for Europeans, with three helpers... There is another agricultural station at Johann-Albrechts-Höhe, in the heights above Zintgraff's earlier Barombi Station which the Government has given up...

During my stay in Cameroon, I have made a close study of cocoa planting... Cocoa is sown in the rainy season. Several beans, usually three, are put in a hole of 2 cm depth. As soon as the plants appear, the weaker ones are removed, and only the stronger ones nursed. The cultivation of plane trees is recommended to protect the young trees from sunburn... The flowering period is in March and April; harvesting takes place from August until December with gleanings in January. A cocoa plantation needs four years before any return can be expected, and six before a bush comes into full fruit. A single bush produces about 20 pods... each containing 40 beans, so a mature bush can produce around 800 beans. Picking the pods requires the greatest care so that the fruiting branch remains undamaged. After that comes the opening of the pods, then the cleaning of the beans and their three days' fermentation when they turn from a rosy violet to chocolate-brown. Then they are cleaned again, dried, preferably in the full sun, and are then ready for export . . .

Only Negroes are used as labour, some locally obtained, some of them Kru and Palmas folk from Liberia. The labour force from local sources is not very well suited to the work so that the acquisition of a labour force is of importance, given that the Liberian labour commands high wages: 10–18 Marks a month and free food. Even if this is paid in goods, there are many extra charges to be taken into account. An attempt to solve the urgent labour question has been the reason for my expedition from Mt Cameroon to Bali.

Up to now [1896] four private plantation companies have been started. At Bonge, 30 metres above sea-level, is the 25 hectare plantation owned by the Swedish trading company of Herr Knutson.[6] It has unfortunately been established on poor lateritic soil in order to be close to the seaboard . . . This kind of mistake is unlikely to recur. The oldest plantation on the mountainside, and with 216 hectares, the largest, is that at Bimbia or Man o' War Bay . . . Its owners are the *Kamerun Land- und Plantagengesellschaft* (KLPG), mainly composed of Hamburg traders. The company began in 1884 with tobacco but since 1886 has turned to cocoa. It was started under the excellent leadership of Herr Theuss and is now worthily run by Herr Friderici. A special advantage is its nearness to Man o' War Bay where large ships can anchor, and it is thoroughly equipped with modern devices – a light railway and up-to-date drying, washing and fermentation facilities. At present, 162 hectares are devoted to cocoa, $8\frac{1}{2}$ to coffee, and 3 to garden fruit and field crops. In 1889 the M'Bamba plantation, a couple of hours from the bay . . . was acquired and planted to about 50 hectares of cocoa and five of Liberian coffee. It is run by Herr Rehbein.

On the west of Mt Cameroon, and close to the sea, lies the Bibundi plantation, on very good soil. Bibundi has the highest recorded rainfall in Africa, experiencing occasional showers even in the heart of the dry season . . . It produces good tobacco similar to the Sumatra type in aroma, but its manager has recently turned to cocoa and during my stay has planted some 125 hectares with cocoa . . . The Company has since expanded by taking over plantation lands at Isongo Udje and Mokundange. It is thoroughly well equipped with light railways and modern machinery. Apart from three Europeans it employs 57 Bakokos, from the south of the colony, 21 Dahomey women,[7] 92 Liberians, and some five Hausa and Gold Coasters. The fourth plantation, Debundscha, lies to the south of it. It was established in 1889 and is doing well.

According to Professor Wohltmann's detailed studies, which are confirmed by my own observations, there are thousands more hectares of conveniently located land on the south and west side of the Mountain available.

It might be some years before these lands will be acquired and placed under cultivation: the capital for development will be easily raised once it is realised that money can be safely and profitably invested. This would offer no competition to German agriculture but, rather, serve to make Germany less dependent on foreign imports... Professor Wohltmann is at present engaged in setting up a colonial school of tropical agriculture in south Germany, a most desirable development.[8]

Notes

1. Esser, like other travellers, who described the peoples of Cameroon, divides them into two, later three, groups – Bantu, Sudanic, Pygmy – a predisposition influenced by Enlightenment systematics and taxonomy (Linnaeus, Blumenbach, Lavater, etc.), and informed by classical Grecian conceptions of slender athletic beauty. Morgen (1893, 39) is more explicit than most in writing: 'The further from the coast one travels the more beautiful the people become: to put it at its crudest the more they look like human beings. This became even more striking when I encountered Sudanic peoples in the interior.' For an examination of the 'common knowledge' and predisposing myths influencing early colonial descriptions of physical appearance see A. Bouba (1996) in *Paideuma* 42, 63–83, which refers to the non-Moslem groups of the North of Cameroon, and G. Jahoda (1998).

2. Esser, like some other contemporaries including Zintgraff, cannot resist repeated tilts against the Duala: von Puttkamer also thought them indolent and parasitic. For another view of their entrepreneurial adaptability see Clarence-Smith (1993) 195–196, and Yvette D. Monga (1996) in Clarence-Smith ed. *Cocoa Pioneer Fronts since 1800*, 119–136. For German-Duala relations see A. Rüger (1968) in ed. Stoecker, Vol. 2, 183 ff., the many publications of R.A. Austen, for example in the *International Journal of African Historical Studies* 16 (1), 1983, 1–24 and in eds. Fowler and Zeitlyn (1996), *African Crossroads*, 63–80. See also Andreas Eckert's study of Duala protonationalism to 1939 with an introduction by Austen (1997), and Austen and Derrick (1999), a masterly work.

3. In January 1896 the European population in the previous census was officially recorded as comprising 157 Germans, 33 British, 17 Americans (mainly members of the American Presbyterian Mission in the south of the colony), 15 Swedes, 3 Russians, and one each of Australians, Austrians, Belgians, Spaniards, and Swiss; see Rudolf Fitzner (1896) *Kolonial-Handbuch*, 93 ff. Of these only 11 were then wholly engaged in plantations and horticulture as against 70 classified as traders and 42 as missionaries and their families. Distributed among these was a handful of machinists and artisans.

4. The reference is to Professor Ferdinand Wohltmann's *Der Plantagenbau und seine Zukunft*, 1896, on which Esser relies heavily, and sometimes word for word. After practical farming experience Wohltmann studied at Halle, Berlin and Heidelberg universities and then worked on the staff of the Halle Agricultural Institute before becoming a Professor at Breslau, then at Bonn in 1894, returning to Halle in 1905. He made numerous overseas study-trips including one to Cameroon in 1889/90, a brief visit in 1896, and to Togo and Cameroon in 1899/1900. With Professor Otto Warburg, author of many works on tropical botany, he co-edited the influential journal *Der Tropenpflanzer* (Berlin) from 1897 onwards.

5. In the event it was eventually held to be too damp and overcast, especially on the west flank of the range, for the reliably successful production of cocoa, which was prone to attack by fungal diseases. See Kemner (1937), 209–210. Cocoa production moved to other areas, and attracted new entrepreneurs with lower costs of production such as the Kosi. For its distribution in the mid-1960s see J. Champaud, L'Économie Cacaoyère du Cameroun, *Cahiers ORSTOM*, Séries Sci. Hum. III (3), 1966, 105–125, and for its history 1914–c.1960 see Andreas Eckert in Clarence-Smith (ed.) *Cocoa Pioneer Fronts*, 1996, 137–153, and Eckert (1997) for a historical overview of the labour costs of Duala and Bamileke planters.

For brief references to the initial caution of German investors in support of plantations, a long-term investment, see e.g. W.O. Henderson (1962) *Studies in German Colonial History*, 58–9 and W. Kemner (1922), 5. Clarence-Smith (1993) has argued that the plantation industry in Cameroon was never hugely profitable. See also Clarence-Smith and Ruf (1996, 13–15) and Clarence-Smith (1993) for some of the reasons why estate cocoa production was essentially unprofitable.

The 'empty desert' reference here echoes some of the opposition parties' initial objections to the acquisition of Colonies as a useless and unnecessary expense. Bamberger of the Liberal (*Fortschritt*) Party is often quoted as saying in 1885 that the English would be happy 'to see Germany getting stuck in some desert bay',(Koschitzky, 1887, 165).

6. The Swedish company referred to went under the name of Knutson, Valdau and Heilborn's Afrikanska Handelsaktiebolaget. It then had a

central 'factory' at Duala with branches at Bonge, Bavo and Lobe (Ndobe), all on the Meme river, and at Bioki, Moko, Rio del Rey and Ndian. There are numerous references to them in other works e.g. Zöller (1885) 179 ff., and Zintgraff (1895) 65–66, and see also *DKB*, 1890, 25–26. There are several contributions by Valdau and his companions to the Swedish geographical journal *Ymer* and one by Valdau describing his and Knutson's trading journey to the Bakundu and the Barombi lake in *Deutsche Geographische Blätter*, 9, 1886, 30–48, 120–141. In 1890 the DKZ featured extracts from the *Ymer* articles, which dealt with villages along the Meme, developments after 1886 and touch on the 'labour question'. For Valdau's part in the Buea land sale to von Soden see Ardener (1996), 132–136. Knutson's memoirs of his stay between 1883 and 1895, which he had sent to old Bakweri friends, has been recovered intact, and will be published in this series, edited by Shirley Ardener.

7. It could be presumed that these were among the women handed over as perpetual penal labourers to the plantations after the mutiny of the Dahomeyan slave-soldiers in December 1893 was suppressed. For a full account of their gruesome treatment, which shocked German public opinion, see A. Rüger (1960), in Stoecker (ed.), Vol. 1, 97–147, who gives full references to archival and press sources and interventions in the *Reichstag* from a wide range of parties.

8. A colonial training school for planters, farmers and prospective overseas settlers was eventually opened in 1899 in a romantic setting at Witzenhausen near Kassel, with a broad curriculum, with Wohltmann as a teacher but not as director. The curriculum and ideological aims of the school are quite fully set out in the *Deutsches Kolonial-Lexikon*, ed. Schnee, 1920, Vol. 3. It owed its foundation to the *Rheinischer Verband des Evangelischen Afrika-Vereins*.

6

In Cameroon

[Having set the scene for his readers, Esser next describes the party's voyage on the steamer Nachtigal from Sao Thomé to Cameroon.]

The German steamer 'Nachtigal', named after our brilliant Africanist researcher, came to pick us up from Sao Thomé . . . We rode to the port on horseback . . . Hoesch's arm still in a sling . . .

Once aboard we learned of the deaths of four Europeans during the last week – those of the Nursing Sister Bertha, of an English missionary, of a Swedish ship's Captain, and that of the worthy teacher Cristaller, a native of Württemberg. All had died of blackwater fever.

The steamer 'Nachtigal' was assigned to the Governor, and we were to stay in his saloon and bedrooms. Thus, we expected some comfort aboard. Soon we realized that no comfort at all was available, and the Kiel-built steam-vessel was a monster built on the cheap. It was neither an ocean-going ship nor a river-boat, be it steamboat or barge. It was built with several incompatible ends in view – as a Governor's yacht, a hospital ship, a troop ship, a survey ship and, at need, a gunboat. Quite apart from the design problems these different purposes created . . . the paltry sum of 250,000 Marks had been allocated to its construction. It bucketed about fearsomely on the high seas . . . and I felt much sympathy for the fever-struck officials who were sent on a sea trip to recover their health . . . Its maintenance costs, some 80,000 Marks a year, were of the same order as those of a conventional steamer.

We anchored at Bibundi . . . The landscape was reminiscent of Sao Thomé, but on an altogether more majestic scale . . . We were struck by the contrast between the snow-capped peak of 4,000 metres . . . and the many colours of the forest that reached the sea.

In Bibundi we were on German soil and in that protectorate which in my view was the finest and had the greatest future of all despite being reckoned the least known and climatically the most dangerous. Herr Rackow, the plantation manager, asked us to breakfast. We enjoyed the fine Bibundi cigars while listening to the manager's account of why the plantation was moving over to cocoa.

From Bibundi the steamer dropped us off at Victoria.[1] The Government had assigned us to the old English consular house on Mondoleh Island, but since we would have wasted time making the daily passage ashore, we accepted the hospitality of Dr Preuss, director of the Botanic Garden. The practical Dr Preuss rigged up a spacious camping site for us in his dining room and made us feel thoroughly at home.

Our aim was to establish a large cocoa plantation in Victoria. On the Government's side we had been told that an area of some 5,000 hectares with 8 km of sea front, which Dr Preuss had confirmed was first-class land, would be at our disposal. Naturally our first task was to examine the concession.

A hard walk, with much scrambling over lava reefs, brought us along the sea to the site. Dr Preuss kindly accompanied us and we were joined by First-Lieutenant von Besser, then in the service of the Foreign Office, who was to draw the boundaries of our concession. We derived helpful information from both men. Both soil and

Figure 6.1: The much-abused steamship *Nachtigal*

water sources were excellent. Of course there were no paths, and we made our way with the help of compasses and the accompanying Negroes, who skilfully cut away the undergrowth. Our picnic consisted of the curious combination of tinned sardines in oil, pineapples and bananas . . .

The evenings . . . were quite magical with the chirping of cicadas and the bright fireflies in the grass whose light was so bright that one could still read even some steps away from them . . . The massive cliffs appeared dark against a star-strewn firmament. The flickering lights of the native fishing boats, the evening sea breeze and the distant murmur of the waves, added to our enchantment . . . One can readily understand why those who have come to Africa are drawn back to it, despite its dangers.

Our time was spent in study, which included that of existing plantations. Helpful information was given by all of their staff and in particular by Herr Friderici, head of the KLPG's Bimbia plantation.

In the Chancellor of Cameroon, the State-Councillor Dr Seitz,[2] I met an excellent civil servant. A Baden man by birth, he was greatly esteemed locally as an able, energetic, and humane man. My dealings with him led to the increase of the concession by 2,000 hectares at Ngeme and the inclusion of Mondoleh Island. We were of course interested in the inhabitants of the mountains. The Duala, born traders as I already have described, are unlikely to give up their lucrative position as middlemen between the Europeans and the hinterlanders without compulsion. Since they are averse to hard physical labour and only engage in it to satisfy essential needs, any attempt to introduce rational cultivation among them is nearly always in vain. Like other Blacks, they are very dressy and bedizen themselves very tastefully with beads and bright cloth, and are tattooed in undeniably pretty patterns. They live in small palm roofed huts . . . They are also exponents of the widespread drum language with which messages and, indeed, quite long reports, can be disseminated over several miles . . . The knowledge of the code is kept from women, who do not seem to be regarded as fully human but are nevertheless treasured, since wives and daughters are viewed as wealth. Many a poor devil works for years to be able to buy a wife from his chief. Once the man has obtained a wife, the latter has to work for him, and the more she is able to work, the more value she has. No husband makes a fuss if another man makes an offer for his wife, but strikes the best bargain he can and gets rid of her. Girls here ripen soon and are considered ready to be married at nine.

Adultery by women is severely punished, not on moral grounds, but because they lose value thereby.

The Duala are divided into three classes: free, half-free, and slaves. A freeman must have two free parents; a half-free person is the child of a freeman by a female slave.

Also of interest are the Bakweri, living on the mountain slopes and chiefly engaged in cattle-keeping, cattle having become a means of exchange and barter. Both sexes may go completely naked or wear a loincloth. I was often struck by their handsome forms. They are adroit at any sort of labour and intellectually well endowed. They are also known for their boldness and courage, characteristics displayed in their banditries. They have been bitter enemies to the Germans in their battles on the Cameroon mountains.[3] Now they have lost in numbers and influence: of the 2,000 who formerly inhabited Buea there are only some 800 left.

The Kru and Liberians are important in Victoria and its neighbourhood. As a tribal mark they all sport a black tattoo running down the forehead to the nose. They are real workers, physically efficient, and the Kru are particularly skilful and bold boatmen. They know how to manage their boats in the heaviest breakers. The traffic between ships at anchor and the shore is in their hands.

As my inquiries amply confirm, the Labour Question is of the greatest importance for a plantation enterprise. The need to solve this question reinforced my decision to explore the hinterland. I will give the details later on.

As a matter of course I have acquainted myself with what is a reasonable mode of life in the colonies. To maintain good health it is most important to maintain a vegetable garden and keep poultry and stock. It seems odd to me that Europeans in the tropics have not managed to put down more than a bed of radishes, when all sorts of vegetables do so well. I recall the well laid out vegetable gardens of Herr Spengler in Sao Thomé and that of Dr Preuss in Victoria, which supply healthful fresh vegetables. Most Europeans depend on tasteless tinned vegetables as well as tinned meat. Likewise over-indulgence in beer is harmful. Export beer has a strong alcoholic content and further fermentation occurs in the course of its transport in high temperatures. A cheaper and healthier alternative is the light red wine exported from Lisbon at about 50 Pf. a litre.

I was able to get interesting figures on imports and exports. In the year 1895/6 the chief export goods, by value, were:

	Marks
3,600,139 litres Palm Oil	1,391,048
5,960,399 kg Palm Kernels	1,286,356
448,883 kg Wild Rubber	1,469,532
30,484 kg Ivory	391,042
507,041 kg Ebony	76,763
110,905 kg Cocoa	138,239

In the same period, there were 4.6 million Marks worth of imports, the main items being salt, tobacco, rice, iron goods, machinery, glassware, soap, haberdashery, brass goods, and other oddments. Naturally, the order of magnitude of these figures will vary from year to year. The reasons for the fall in wild rubber exports have already been described, but since then the good Dr Preuss has been experimenting with *Kickxia africana*. Generally the exports of ivory are down although elephants are still common in the hinterland. But the demand for ivory is strong and if protective measures are not soon taken the elephant will gradually disappear. It has been estimated that to cover the European demand for ivory alone, 65,000 African elephants would have to be killed. In the last decades, on the average, England alone has imported:

	Th. Kilograms
1840–50	294
1850–60	474
1860–70	556
1870–80	597
1880–90	670

African ivory is held to be superior to the Indian; the tusks are heavier and larger. Likewise, West African is preferred to East African ivory in being slenderer and more transparent, while East African tusks are more twisted and not of such a good colour. All attempts to tame the African elephant have so far come to nothing, alas, and it remains to be seen whether the pessimists, who think it untameable, are right . . .

> [In the next seven pages Esser treats us to the then popularly accepted views of the human and crop colonisation of Africa in which there is a passing mention of the agricultural tasks of women, African food preferences, and yet more on the introduction of New World plants by the Portuguese – maize, cassava, various kinds of pepper, groundnuts, sweet potatoes, tobacco and lastly cinchona and cocoa. What, he wonders, were the African staples before the introduction of these crops?]

Cocoa, this most important product, came to Africa in 1824 and was tried out in various places, but only did well in the islands of the Gulf of Guinea, below the Cameroon mountain[4]... It is impossible to engage the coastal population as plantation workers. For centuries they have lived from trade ... One can hardly imagine the unbelievable indolence of the Duala and their distaste for hard work. Other African coastal groups have initially provided the necessary labour, though at high costs. They are recruited from the Vai [Liberian] and Accra districts ... The costs, which involve commission, transport and the duty to ship labourers home amount to some 50 Marks each; however, the cost of food and personal wages, some 20–25 Marks a month, is not excessive.

One hectare with 500 cocoa bushes to be planted in a year needs two labourers. One can work out from that that 400 workers are needed, under six Europeans, to plant some 100,000 trees a year. If nowadays Black workers earn 300 Marks per annum and White overseers 3,000, planting one cocoa bush costs about 1.40 Marks.

The Cameroon administration now appreciates the situation and seeks to persuade the natives to work above what is required for mere subsistence. The efforts of the administration are generally acknowledged and I can confirm its good intentions in this regard. Worth mentioning too, were its praiseworthy attempts to bring the populous inland tribes into greater contact with the coast.[5] To win over members of the hinterland tribes to employment in coastal agriculture would have the advantage that high commissions and transport costs would be sharply reduced. The notion of setting up a monopoly, which the Government should put in hand, to hire labour from the hinterland to the plantations and to the Government, and to see to it that they were regularly paid and fed so that they could return home at the end of their contracts in good health, had occurred to me.

If such were the case the entrepreneur would have a guarantee that he could always obtain the labour he needed, while the worker would be provided with the certainty of good food, regular wages and a timely discharge to his own homeland. But for this to happen the administration would have to have agents in the hinterland and I think that the hinterland chiefs – of course after careful selection – could be called on to act as such. Therefore, the purpose of the expedition to Bali was to arrive at such an agreement with a hinterland chief, and thereby to contribute to the solution of this urgent question. It had been discussed at length with Consul Spengler who also gave us superior information on the Portuguese role in the enrichment of Africa with useful plants.

Notes

1. Wilhelm Kemner, Esser's successor as managing director of the WAPV in 1909, recorded in his pamphlet *Was wir verloren haben* (1922), 5, that the Station Leader von Besser was told off to meet the three visitors. Von Besser recorded the meeting in his diary as follows:
   'The three gentlemen seemed very ready for the fray in their yellow boots reaching to their thighs, huge grey felt hats, buckled-on bush-knives and tall bamboo walking staves: they looked as if they were going to take on the whole of Cameroon. This get-up of theirs, pretty impractical in the forest region, had been procured on the advice of Zintgraff who, theatrically, preferred it because it made a great impression on the natives.'
   Zintgraff's earlier expedition costume is illustrated in the frontispiece to *Nord-Kamerun* and is reproduced at Figure 7.1.

2. The reference is to Dr. Theodor Seitz who later returned to Cameroon as Governor, 1907–1910, and was subsequently appointed to be Governor of German South-West Africa. His policies as Governor in Cameroon reflected those of Dernburg, Director of the newly formed *Reichskolonialamt*, and are quite lengthily discussed by Rudin (1938), Hausen (1970) and in the second volume (1929) of Seitz's memoirs. The first volume refers to his earlier experiences in Cameroon. In 1920 he became President and later an honorary president of the German Colonial Society (DKG).

3. The Bakweri resistance is described, with full contemporary references, by E. Ardener (1996), 41–150.

4. The mission-introduced cocoa of the migrant Gold Coast farmers receives no mention by Esser at this stage. More surprising is the omission of the 'Victorians', the Baptist colony, initially of 'Liberated Africans' who were refugees from Fernando Po, whose small-scale production of 'trade cocoa' had preceded the German annexation, and was followed by the larger scale production of the Duala. See R.F. Burton (1862), Vol. 2, 69; Wirz (1972), 203–204; Clarence-Smith (1993), 194–196; and a pseudonymous paper by 'Merkur' in *Koloniale Rundschau*, 5, 1912, 268–272.

5. The good offices of the Government in labour-supply included the provision, after 1893, and increasingly after 1899, of penal labour, either prisoners of war, women included, exacted from groups resisting the German advance as a war-indemnity, or provided willingly or otherwise by Chiefs. See von Puttkamer (1912), 105 and Rüger (1960), 195–197. The KLPG manager, Friderici, had even proposed that, since training was wasted on short-stay labourers and involved expenditure on European overseers, penal labourers should be retained for 5 years. See his

letter originally published in a Hamburg newspaper and reproduced in *DKZ* XI, 1898, 162–163. Other devices mooted included a general system of forced labour (*DKZ* X, 1897, 23). News of the shocking mortality among hinterlanders reaching the plantations had begun to stiffen resistance to recruitment.

7

The Expedition to Bali

So we set off for Zintgraff's old blood-brother and ally Garega,[1] king of Bali. Naturally the contracts we intended him to sign were to be formulated in such a way that they did not conduce to the circumstances to be found in Sao Thomé and Principe, where the ancient institution of slavery continued under the specious label of 'contract workers' [*Kontraktierte*], which sounds respectable enough. Undisturbed by the warships of European nations cruising the Atlantic Ocean, large cargoes of these poor creatures were disembarked every year on these islands from Novo Redondo, Angola's notorious transshipment port, where they had signed contracts for a term of five years.[2]

The notion of time as a calculable duration is completely unknown to the Black man. He has not the slightest notion what 'five years' are and what he should understand by that. He just signs a contract for the term of 'five years', and when these have elapsed, the owner of the plantation can easily tell him: 'You have worked only five months for me'. And one thing is certain, only a small percentage of the unfortunates who have been shipped to the islands will ever leave them again. On the Victoria beach, numerous canoes can be seen in which these people, driven to despair, have tried to reach the secure mainland from the island of Fernando Po in order to live and breathe as free persons under the protection of the German government. A look at the surging sea at Victoria is enough to make one realize how many of these canoes were sunk in the play of the waves, unable to reach the saving shore with their human crew. Today, it is fortunately impossible at Victoria to establish this sort of business in slaves – a mockery of civilization.[3]

In Africa, before being able to start off, every traveller must be ready to overcome numerous difficulties and contend with annoyances of every kind. We, too, had to face them. First the Woermann steamer which was supposed to bring our expedition equipment from Hamburg to Victoria, arrived five days later than scheduled, and when the eagerly awaited ship finally arrived we learnt with irritation that the tents which we had ordered from a Berlin company had been left behind. We were supposed to await the next steamer which was expected only about four weeks later. Since the rainy season had already started and although the absence of the tents would bring troubles in its wake, we nevertheless decided not to wait but to set off. If we had waited any longer the rivers would have risen so high that crossing them, and the entire journey, would have become impossible.

Our expedition proceeded, especially on the way there, mainly along known paths and with more or less accurate maps at our disposal. Nevertheless, various parts of the route had to be re-explored. In Central Africa, as soon as European influence ceases to be tangible in a certain area, such a shift in conditions may take place within as little as four years that the area may no longer correspond with earlier descriptions.[4] Bloody wars may have raged and villages and towns destroyed or else the very chiefs, who had enjoyed most prestige under the protection of the Europeans, may, after their exit, have lost power and territory, and given way to more recent and more powerful rulers. Those acquainted with African matters will be aware how often Negro peoples tend to reposition their villages for one reason or another, perhaps due to superstition, or an epidemic, or because of the increase in elephants and similar plagues. And then the decision to change their place of residence is fairly easy since a move does not involve many difficulties. After all, the move is pretty simple since it is confined to the transport of some pots, fowls and goats, which usually comprise the entire possessions of these people, while the few simple mud- or palm-rib huts can be constructed anew and protected by palisades at the new site within a matter of one or two weeks.

We did not find a single trace of some sites which were indicated as larger villages on the maps we had. On the other hand, many large villages, not recorded on the maps, had emerged in the interim. Where we had expected, on the basis of earlier experience to be welcomed in friendly fashion, people received us with scepticism and reserve. Where we feared, however, to encounter hostility and difficulties, we were received hospitably. This change in only four years might have been accelerated by the fact that the natives believed that once Zintgraff's stations had been closed, no White man would ever again set foot on the route to the Grassfields.

I should record that on our march to Bali the four stations given up by the Government were found in ruins. The reason for closing down the stations has been a matter of debate. As for me, I can only approve of this step.

However right it was at the beginning of our colonial endeavours to throw light on the interior of our colonies and to mark their inland boundaries by treaty, it was wrong, in my opinion, to seek to control the entire interior at the same time and to set up expensive as well as useless stations in the hinterland.

It did not take the Government long to come to the conclusion that it had to leave the interior to fend for itself, and to start to devote its energies to the exploitation of the coastal strip by extensive cultivation, and to support the establishment of plantations. In this respect, the Government has achieved considerable success in a relatively short time by setting up the experimental plantations and stations at the coast we have already described.

The four abandoned stations on the Bali road would have cost at least 20,000 Marks each, on a conservative estimate, in all 80,000 Marks a year. The entire value of the products, carried by the natives down that road to Mundame amounted, I have ascertained, to at most 25,000 Marks a year; and the most curious thing is that despite the closing down of the stations this trade has not decreased but, on the contrary, has even increased a little. This proves that a large number of stations was not necessary once the road had been opened by expeditions. In this respect, too, we should learn from the Portuguese, who have concentrated all their power at the coast, for as long as larger means were not affordable.[5]

We took the groggy pinnace belonging to His Royal Highness, the Prince Manga Bell, King Bell's first son, up the river Mungo to reach Mundame by the river route. We were very glad to have chartered this aged pinnace, even though it needed a two-hour rest after the same amount of travelling time, since an epidemic of malfunctions had struck the Cameroonian vessels at that time. The Government's steam-vessel had sprung a serious leak after a collision on its way to Edea, Woermann's No. 1 was being repaired and No. 2 had run aground just as it put to sea to carry us. The Jantzen and Thormählen vessel was not fit for service as a result of a defect in the cooling pump, and so, apart from Bell's old pinnace, there was no craft at hand to take us up the river.

I would like to mention one incident as evidence of the Negro's incredible unreliability and lack of a sense of obligation. We had made a contract with the people of Bimbia according to which they would take us, some days later, with three war canoes of 20 rowers each, up river to the Mungo villages. As to price we had agreed that for each rower 10 Marks and for each canoe 25 Marks were to be paid to them, 675 Marks in all. At the arranged time the 60 men appeared with their canoes; their leader, however, declared that he now required 25 Marks for each man – that is, instead of the agreed 675 Marks, 1575 Marks (!). I felt strongly inclined to force these people, who had attempted to exploit our plight, to carry out their duty. However, I refrained from drastic disciplinary action, refused their services and confined myself to lodging a complaint against them with the Government. The defaulting natives were later called to account by Dr Preuss, the acting District Officer.

The Mungo, Cameroon's most beautiful river, is navigable up to Mundame; from there on, unfortunately, insurmountable rapids set a permanent barrier to all forward motion.

The banks of the Mungo are magnificently covered with forests; sometimes one crosses dense virgin forest made up of hundred-year-old giants, splendid specimens, which could barely be encircled by thirty men; then come vast palm groves with their characteristic charm for Europeans.

And everything here teems with life. One can see sea eagles, herons, snakes, and monkeys, as well as multicoloured parrots on the trees, while on the surface of the water there dance butterflies and dragonflies the size of sparrows. Now and then one hears the trumpeting of elephants, the cry of predators, and the melancholy and monotonous honking of the iguana. And then a huge uprooted palm tree straddles half the river and iridescent fish splash around in safety under its magnificent fan-shaped leaves.

The paths cleared by elephants or natives through the palm groves and virgin forests look rather like mining tunnels, enveloped as they are by thick leafage.

About 35 km from the mouth of the river, the virgin forest begins to thin out, and one sees huts and villages, farms of plantains, cocoyam, corn and sugar cane as well as the numerous dugout canoes of the natives. And so charming pictures of Culture alternated with the impressive grandeur of Nature.

After two days journey, with new impressions at every turn, we reached Mundame, a place which, until 1892, was a Government station. Today, there is a factory there which, together with the station 'Johann-Albrechts-Höhe', constructed only some months ago and lying 30 km to the north-east, represent the last outposts of European civilisation in the northern protectorate of Cameroon.[6]

Our approach had already become known in Mundame a day ago by means of the so-called drum language. It was from this place that we intended to start our journey into the interior, and here it was that we took our final leave of European civilisation. 150 Bali warriors awaited us, sent by King Garega for our protection after he had learned of our approach.

The people of our expedition consisted, apart from these Bali, of Vai and Kru boys, both from the Negro Republic of Liberia. Despite sharing the same homeland and being tribally related, they are completely different from one another. The Kru boy is perfectly suited to the water and sea; he is a unique rower, but he is timid on land and a bad walker. This was why we decided to dismiss a part of these people in Mundame after we had ended our river journey. The Vai boy, however, in contrast to the Kru, is perfectly created for long marches, a proper *lansquenet* for whom a vagrant life together with wine, women and play means everything. Apart from his happy-go-lucky and imperturbable nature, evident in his indifference to danger, he is, as Dr Zintgraff aptly says, born with the venturesomeness of the gambler. The saying, common today along the entire West Coast and applied whenever there is trouble, derives from him: 'Today be today'. This means that one just has to put up with what cannot be changed. He lives for hunting and war, for gaming and dancing. When the Vai has received his pay he indulges his passions, until the last penny is spent. Then he attaches himself to any expedition, a White man's or a warlike chief's, and faithfully shares all its pains and dangers with no hesitations or grumbles.

It is not to be wondered at that in calm and peaceful times such folk are difficult to employ and become a nuisance in the settlements. Indeed, the police headquarters at Victoria were very pleased that they had got rid of about 50 of the homeless, wandering young fellows we had recruited, albeit only for a few months. All of the Vai wanted to return after the march to the beauties of Victoria and not to accompany us to Mossamedes.

Insofar as the intelligence of the Bali is concerned, I would like to record that I was successful in Mundame in drilling and training 25 of the Bali people in the use of the M 71 rifle in less than six days, and that their marksmanship was excellent. The speed with which they learned, the neat execution of their drill, and the calm and reliability they showed in the course of their shooting practice was quite remarkable. They would make excellent material for our

Schutztruppe if we could succeed in familiarizing them with the permanent constraints of the life of a soldier. They are long-legged sons of the highlands and stand out, insofar as their intelligence is concerned, quite strikingly from most of the other tribes of the protectorate.

However, although they were very fast walkers, they were completely unfamiliar with carrying loads. We therefore decided on a procedure which was certainly a daring experiment, but it proved successful and enhanced the speed of the expedition considerably.

We gave loads in boxes weighing 25 kilos to the Vai boys. All the other loads were unpacked and the contents were divided into small loads of 10 kilos each. We then divided 80 of the Bali into eight platoons of ten men each, gave each platoon a leader to whom we gave ten divided loads for his people. We took down the name of the leader and the contents of the loads allotted to him. Thus, one platoon had trade or barter goods, another tins, a third munitions.

When we were in need of something, we only had to call up the leader of the specific platoon. He then arrived with one of his people and we took out what we wanted, recording what was removed to ensure better control. As a result of having about ten carriers responsible for one type of load, it never happened that the one man who carried the necessary item had not yet arrived.

To the credit of the Bali, I have to declare that not a single item of all the goods given to them loose went astray. They had speedily woven ingenious containers out of the fibres of the palm, *Raphia vinifera,* and in these they carried the carefully packed loads on their heads.

The first day of our march brought us to Ikiliwindi, the largest village of the Bakundu tribe, which had settled north of Mundame. Ikiliwindi has about 300 houses with approximately 1,000 inhabitants.

This was the place at which the earlier research expedition of the explorer Dr Schwarz had broken down – that is, he was forced to return. The chief of the Bakundu received us, in contrast to Dr Schwarz, in the most peaceable manner; he brought us goats and fowls as presents, sold me his very artistic fetish in exchange for an accordion, which, since this new god was able to speak, seemed to him a far more appropriate object of devotion, and he was courteous in every way.

Asked why he had attacked the African explorer Dr Schwarz, he denied ever having waged war against any White man and as proof of this he referred us to the witness of the elders of the village.[7]

The route from Mundame to Ikiliwindi is beautiful and romantic, rich in canyons and rocks. The following day we passed through various slave villages and, after crossing some strongly flowing torrents, came to Baduma. Shortly before reaching this village we crossed a swampy depression, a so-called dancing ground of elephants. The innumerable elephant tracks made progress difficult; the soft ground often made us sink up to our knees into these tracks.[8]

Baduma is governed by Chief Mokurru, the most respected hunter of elephants in the entire area. His huts are surrounded by well over 50 skulls of elephants. He himself, in the company of his council, received us enthroned on these skulls. He was obviously very pleased to have the chance to greet White men, and he brought us and our people many goats, plantains and cuts of elephant meat. The settlement is idyllically situated on the top of a small hill in the virgin forest.

Figure 7.1: Zintgraff in expedition costume, from the frontispiece to *Nord-Kamerun* (see note 1 of chapter 6)

The next day, in Bulo Nguti, we were successful in acquiring, in exchange for goods, three man-high fetishes in good condition as a result of lengthy negotiations with the chief and the sorcerer of the village. They will be of special value since neither I nor other travellers in the protectorate have ever found similar ones. So far as I know, such fetishes are not yet represented in the Berlin *Museum für Völkerkunde*. They include a man and a woman, both life-size and fully naked, and in between them a broad board with carvings: these depicted heads of human beings and animals surrounded by flowers, hearts, squares, balls and bones – signs the interpretation of which I have to leave to the experts.[9]

Worth mentioning too, is the architecture of the Bakundu who build all their villages on the same model.[10] A large, broad street is lined on both sides by impressively large and handsome huts, which are three to four meters high, four to six meters wide, and often ten to fifteen meters long. The huts are built, as is usual in the entire area, of poles and leaves of the *Raphia vinifera* (wine palm) and

Figure 7.2: The 'Fetishes' collected by Esser in Ikiliwindi (the tallest) and Bulo Nguti, as illustrated in Esser's book.

remain wholly dry inside, even in the rainy season. Through the huge main hut which serves as the *foyer* and living area for children, slaves and goats, one enters a small courtyard which is closed off by three smaller huts. In these the family head resides with his favourite wives.

In most of the villages a large communal meeting house (the so-called 'palaver house') is located at its centre; it is built in even larger dimensions than the *foyers* we have described. Here all important decisions are debated and agreed; the drums are stored there and the sorcerer produces his medicines within it.

The question of whether the Bakundu are cannibals or not has been much debated. On the basis of close enquiries I have to answer the question in the affirmative, indeed quite positively. Not only have the peoples of the highlands assured me that this is so, but some of the Bakundu themselves have confessed to it in my presence. My inquiries enable me to make the following so far unrecorded revelations on this question.[11]

The whole large tribe of the Bakundu, which is widespread to the north of the Cameroon coast, has a common secret society, the initiation into which is possible only after a vote and the payment of a fee of ten goats or seven dogs. The large society is divided into smaller associations in the individual villages. At each full moon the members of these smaller associations have meetings, and once a year they all meet for a big feast at a particular site in the forest. Then a big feast takes place, for which oxen, dogs and human beings are cooked together in earthenware pots and eaten in a stew. It is this mixture of human flesh, dogs and ox-meat which is supposed to be particularly aromatic and delicious, a tasty treat for these palates. The high entry fee ensures that only the richer people can afford to become members of this association. As their badge they wear the red feathers of the parrot in their hair, and no Bakundu dares to wear a shirt, or a hat, or carry an umbrella, without being a member of the association. It is striking that these emblems of advanced European civilization, together with the carrying of fly-whisks, have become the insignia of an association serving such barbarous ends.

When a member feels death is near the other members of the association can anticipate the pleasant duty of cutting him up and eating him, so that he is able to live on in this way in his friends and fellow-members and so that no part of his body will rot and perish in the earth. Corresponding to the means of the man concerned and to add prestige to the festivities, a certain number of his slaves are slaughtered and eaten along with him.

When a chief is dying, his butchering has to be postponed until up to 40 or 60 slaves have been rounded up in the neighbouring slave villages. Had the slaughter of the chief taken place too soon all the slaves of the neighbouring villages, on hearing of the chief's death, would have fled into the bush and hidden until the corpse had putrefied.

As we passed by Miyimbi, where the king had died six weeks ago, we were told that thirty slaves had been slaughtered in his honour.

We were often to encounter in the forest the grisly traces of such savagery, human skeletons, mixed up with the skulls and bones of dogs. The heads of the slaughtered human beings were pinned to the outer walls of houses as a sign of wealth and power and we often saw, hanging from the roof of palaver houses, baskets full of skulls blackened by smoke. The only pleasing thing I can remark is that this tribe is despised by all the neighbouring native tribes because of its bestial passions.

If I unroll such a picture of these particular Black Africans, for I haven't found more disgusting habits anywhere else, it is to raise a complaint against the internal slave trade and the capture of people undertaken by the natives among themselves, which is still permitted. I raise this complaint as a matter of the greatest urgency.[12]

The following days involved us in enormous efforts in crossing the strongly rising rivers. It was impossible to cross these torrential masses of water by swimming. So our Blacks dropped into the water in detachments of thirty men, hanging on to each other tightly and striving – a huge, living cluster – to reach the other bank. Many a piece of luggage, of course, was damaged or wholly lost during this exercise.

Generally the Blacks proved to be very skilful in crossing rivers, and it gave them a lot of pleasure each time that we, too, had to work our way to the other bank in the middle of such a cluster. It was then that I was sometimes shocked by the sight of our own bodies, because, among all these deep black, strong and beautiful shapes, we appeared almost abnormally white. The Blacks were at first most amazed that the water gave us White folk the same trouble as it gave them. In their view, the water was the element, the home country, of the White men, believing, as they did, that they came up from the depths of the sea, via the shore, to encounter them. They claimed that the White man was a fish-being, because he had white flesh like that of a fish.

On the fifth day we reached Kombone, still moving through the country of the Bakundu. At this place we had to cross the Mungo by a hanging bridge fabricated by the natives. These bridges are very skilfully made and remind one of the hanging bridges of the Indians. Imagine an enormous hammock made from rattan, but hereabouts belonging to a creeper, which often climbs for more than 100 meters up the giant trunks of the jungle. Across the bottom of this hammock, at intervals of about 10 meters, sticks are placed over a twisted rope of the rattan-like creeper, which runs, as thick as your arm, along it and which is made fast at the sticks: you will now have an incomplete, but approximate picture of these bridges. These are, in every case, attached to two trees standing opposite each other on each bank, and can be reached with the aid of ladders made of the same creeper. These bridges swing alarmingly to and fro, the more so as one reaches the middle; indeed, treading along these ropes, which are often more than 80 meters long, I always had the feeling of being a tightrope walker. One wrong step on these bridges and one is irretrievably lost, because the woven side nets of raffia are mostly so weather-beaten and torn that they would not be capable of holding up a falling body. The purpose of the nets is to diminish, as far as possible, the shaking of the rope that is knotted into the nets.

The bridges are maintained by the village communities and may be crossed by everybody free of charge.[13] It goes without saying that it is impossible for cattle or carts to cross the bridges.

After Kombone, we came to the Batum [Batom][14] tribe. Here the architecture of the houses, as well as the layout of the villages changes. The long streets with huts on both sides are replaced by the compound system. The compounds lie at a distance of three to five minutes from each other; their various living huts are connected to one another in the shape of a large rectangle.[15] In place of the thin-walled raffia-ribbed houses we had seen earlier only clay-walled houses were to be found here because we were already in a cooler clime. Not only the walls but the benches inside the huts are made of clay; these are polished smooth and armrests are added. Often these are painted in patterns in which black, white and blue dots alternate upon a black and white squared background. The entrance to these huts is mostly so narrow that one has to creep into them.

Our path now slowly rose, and we passed for several days through virgin forest, in which we only found scattered houses and so the feeding of our two hundred men often became a problem. Fortunately we succeeded in shooting several elephants, whose meat served to feed our people.

Then we crossed the river of Calabar, which here, in German

territory, is unfortunately not navigable because its rapids are at the German-English borders.[16] The English, since they can ship their products far inland, have successfully constructed huge factories at the border, and are consequently in a position to draw the trade away from the road to Mundame, and so from German into English territory.

The natives save about seven days of travelling on foot if they exchange their palm kernels and their rubber for the guns, cloth, and brass wire already close at hand, instead of carrying them all the way to Mundame; moreover, the English can pay them here, in the interior, almost the same prices as we, the Germans, can pay them in Mundame because one more day of river-transport does not cost them much.

Altogether the march had gone well so far and, as I have already remarked, the expedition was well supplied with meat because of the elephants we had shot. It was often amusing to look on when the Blacks gave each other a belly-massage with their feet, when they had difficulties in digesting the tough meat. One man, nevertheless, died of stomach cramps as a result of overeating elephant meat.

Then we reached the Banyang country. The people distinguished themselves by their hairdo from the Negro tribes we had met so far, in that they plaited their hair into small braids, adorned with beads. Moreover, they always walked around armed with Dane guns and a large machete. Over the shoulder they sling a piece of softened barkcloth, which is used as a sweat-rag. At festivities or in wartime they appear in helmets which are woven of palm fibre and covered with elephant hide; on top is attached a piece of monkey fur.

The next day brought us to Tinto, a village with a marvellous view down the plain behind us right up to the sea and the snow-covered Cameroon mountains. Up here there was formerly a government station which, like all the others, was given up in 1892.[17] The remnants of the buildings were scattered around. From Tinto, the path led across several hanging bridges over the river Fi. We were lucky enough to shoot an antelope. Its meat was very welcome, since a food shortage reigned. The natives, with no need for our European goods at all, turned out to be more and more choosy insofar as goods for exchange were concerned.

We had arrived at the point where tobacco ceased to be of value as a means of exchange, whereas beads were not yet favoured as a means of payment and no great value was attached to cloth. Only gunpowder and salt went down well.

Let me now interpose some words about the actual means of exchange.

In the whole of coastal Cameroon, and for many days' march inland, the most favoured means of exchange is tobacco; it is put on the market in leaves. As the natives do not smoke it, but prepare snuff from it, there is generally a high demand for the Kentucky tobacco because of its spiciness and its large leaves. The natives grind the leaves between stones and then mix the tobacco with the ashes of burnt plantains: this produces a very hot powder. I observed two ways of using it, which may be of interest to European snuff-takers. Most Blacks take snuff by heaping the powder on the palms of their hands and then they slowly inhale it through their nostrils. Others, however, first put the powder on their tongues and after it has given them a sharp burning sensation, they stuff the wet mush deep into their nostrils with their forefingers.[18]

At the coast and in the not too distant inland villages tobacco is the main trade currency; beads replace it in the hinterland of Cameroon far into Adamawa and the Lake Chad region. Tobacco is worthless here, because the peoples of the interior grow it for themselves. But trading with beads is not as easy as trading with tobacco. Each tribe has its own preference in beads of a specific size, shape and colour. If the beads offered do not exactly match their expectations, nothing at all is offered in exchange: they are in fact completely worthless. One also has to distinguish between beads for women and those for men. Among some of the tribes only the women wear the common glass beads, in particular the farther one travels into the interior. The men, on the other hand, demand polished stones.

Like everywhere else in the world, women's whims reign and create the fashions which make their menfolk sigh and suffer. This extends to the poor explorer, deep in the interior of Africa, who has to put up with the women's fancies and the vagaries of Queen Fashion, and that can mean suffering from bitter hunger, just because, all of a sudden, a tribe changes its preferences. The women demand a particular type of bead which the explorer, despite his well-selected sample case, does not happen to have in stock. These beauties cannot be persuaded to wear a string of beads which, in their eyes, has gone out of fashion. They will not give anything for them, and would not even allow you to warm yourself at the fire in their huts or cook food for you. They regard this unmodish traveller with open contempt for his ignorance of their latest fashions, and even deride him when they notice his or his people's discomfiture.[19]

When this happens there is nothing for it but the sacrifice of cloth to these fair ladies; and since they actually wish to dress themselves *only* in your beads, your cloth is of little value to them and has to be exchanged for trifles in comparatively large quantities.

In Miyimbi we allowed ourselves a first day of rest. Only a few weeks back, as I have already mentioned, the local chief had died; on the occasion of his funeral celebration thirty slaves had been killed.

In answer to our inquiries the natives denied having eaten the slaves; they had only drunk a little of their blood during the sacrifice. This appeared plausible to me, since the Banyang do not have the reputation of being cannibals.

The deceased king, Difang Tabe, a notorious tyrant, was so feared that even after his death the superstitious natives ascribed a tornado, which had raged there a few days before our arrival, and an elephant, which had broken into the farms and done a lot of damage, to their deceased sovereign. He had raged through the village first in the form of a thunderstorm and then in that of an elephant.[20] And why?: because the thirty sacrificed slaves were not enough.

And now a boy of sixteen years, called Ascheng, a stupid and inexperienced child, is reigning there. He is unable to rule these easily-panicked natives with a rod of iron and keep them together, so a good number of them have already moved out, and the rest of them were seriously considering whether to move the entire village in order to escape from the vindictive spirit of the former king.

An exhausting march of 45 kilometres led us across a dreadfully swampy forest to Sabi. The abundance of game is extraordinary in this area. We came upon the rarest wildfowl, dozens of monkeys, large herds of antelopes, and even of buffaloes, not to speak of many elephants.

From Sabi onwards the gradient of the route became steeper and steeper. The route led us, with many twists and turns, from hill to hill, across pebble beds and torrents, up to Ashu, at a height of 800 meters. Again and again I had to admire the Negroes who were able to walk barefoot on their hardened soles over the sharp rocks and stones whereas the almost new leather soles of our boots had already begun to show signs of serious damage.

The African nature about us was marvellous. If all the mountains up to the height of 1000 meters had not been thickly covered with the most beautiful palm groves as far as the eye could see, one could imagine oneself climbing the mountains of Switzerland or Tyrol. Those palm trees, however, always remind you of where you are.

All of the palm kernels they bear are carried to the English factories at Bakun near the German border. The entire quantity of palm oil to which the trade of the English protectorate, centred on the Calabar [i.e. Cross] river, owes its prosperity, originates here and gave the Cross River its other name, the Oil River [as all the navigable rivers between the Niger and the Cross were usually called].

And torrents run like silver strings from the summits through the forests, producing gorgeous waterfalls here and there. The whole of nature becomes stranger and stranger and with every step more of the landscape comes into view from the heights. The wild rugged mountains rise, as I said, to a height of 1100 meters covered in a thick mantle of palm trees, and beyond them bald rocks jut into the blue sky, covered here and there with scanty grass. And now a fresher air wafted around us; the stifling heat of the bush had disappeared. We are approaching the grassfields.

At Ashu we had a rest and relished spicy palm wine served to us in ornate calabashes: it is with good reason the favourite drink of the mountain folk. Its production is very easy. They bore a hole into the young wood of the palm tree (*Raphia vinifera*), which grows exclusively in West Africa, and insert a pipe into the heart of the wood which directs the generously flowing sap into a calabash which is placed underneath it. As with the juice of the grape, here, too, one can speak of different site conditions, for it is not only the type and age of a palm tree which is decisive for the quality of the product, but also the soil and location in which it is grown. King Garega, for example, was able to distinguish precisely between different vintages and varieties. Palm trees should only be bored once in every two years, else they die. The fermentation is very rapid, and after two days the sap has become quite inebriating. The Bali man always consumes this wine while chewing kola nuts, which grow excellently here.[21]

The kola nut is an excellent antidote to the effects of the alcohol, a fact which might not yet be very well-known. Although I saw Bali people and even King Garega get through up to 10 litres of this spicy stuff in a meeting, I have never seen any sign of drunkenness among them. The Bali claim not to be able to drink half of the quantity without the kola. *Probatum est!* The kola nut, also called 'gura' and 'ombeme', is the fruit of the *Stinkbaum* and in my opinion could be grown in plantations.

The kola nut contains:
2.0% of Theein and Theobromin
42% of nitrogen-free extracts
8.2% of proteins
according to the latest analyses.

We took some of the finest species of the area back for the Botanic Garden in Victoria. They have survived and prospered beyond all expectations.

After having enjoyed the palm wine and segments of kola nuts, the grassfields, for which we had waited many a long week, now beckoned seductively to us from the heights of Babessong. The mere thought of being so close to our long-awaited destination had such an electrifying effect on our people that, tired as they were and despite the strains of the day, all strove to climb this last steep rise and reach the top before sundown. The Bali tore in front singing loud battle songs. They were in the highest spirits, being near their homeland at last.

Anyone who has climbed the Feldberg in the Black Forest, looked down from its grassy ridge onto the wooded fir-crowned slopes, and has let his gaze wander further along the deep-cut, broad, fertile valleys, can't help thinking of the Feldberg when climbing the heights of Babessong. Not only are the formations similar, but we encounter the same numerous torrents with their charming waterfalls. But here, instead of the dark fir-woods we see the light-leaved palm trees, instead of fertile valleys and open fields below, only an impenetrable swampy jungle.

At dusk we reached the entry to the grassfields. The setting sun smiled on us as if congratulating us on our arrival, gilding all of us and the entire landscape with its last rays.

Now we were standing at 1,450m in the village of Babessong. Here, the architecture of the village differed once more from that which we had already encountered. Here we found huts which were small rectangular boxes of clay with a high pyramid roof, and very smoky they were too, because the native has to burn a fire day and night to face the cold up here, which he feels keenly.

Hundreds of warriors, armed with spears, awaited us in a squatting position, holding the spears upright in front of them; they looked at us tensely and then at their chief; and only seemed to calm down when they saw us negotiating peacefully with their ruler.

Let me remark that Babessong is a pig-breeding country, a fact which we found very agreeable, for the chief did not give us goats which we had had almost every day for weeks, but a present of three giant pigs. On that very evening these vanished without trace into the hungry stomachs of our people.

And next morning when I stepped out of my hut, what a view! In a swirling mist below lay valleys and groves, and up above, in brightest sunshine, lay the far-spreading grassfields. Along with this there was the most splendid mountain air which our lungs breathed in with delight. In the distance I saw antelopes and grazing buffaloes, and above us glided mighty eagles.

Let's go on to Bali. The destination of the march was a thought in everybody's mind: in half an hour everybody was ready to go.

Nobody was walking any longer; the Bali stormed ahead like young hounds. The Vai boys did not want to be outdone and along the paths of the grassfields, which were very easy going compared to the difficult marches in the forest, a regular race soon developed. Everybody panted under the loads, but the people stormed on further and further continuously singing battle songs. From time to time one of them put down his load and jumped rejoicing into the head-high grass, performing a wild war dance. We were all happy and in high spirits. The fresh air had a refreshing and stimulating effect on our lungs which had been deprived of this element.

We had to climb over several mountains about 1700 to 1800 meters high, then the path led us gently down again to 1420 meters, the approximate altitude of Bali. Steep gorges and torrents had been bridged by the Bali by throwing a log over them. With a good head for heights the Blacks walked over these logs, whereas we preferred to crawl on all fours over them. At about half way to Bali, we met the first messengers of Garega, the chief of Bali. They delivered the warm words of welcome of their king and escorted us to the borders of his realm.

Here a solemn delegation awaited us and presented us with a breakfast in clean baskets which taught us to respect the king's culinary good taste. There appeared four spit-roasted capons – the Bali rear capons – warm corn bread, a dish called 'Essuga', made of local vegetables, which were neatly wrapped up in banana leaves, and a porridge of barley [sic]. Then out of a huge raffia sack was produced a skilfully ornamented calabash with burnt-in arabesques, full of palm wine; the obligatory kola nuts were not lacking either.

The Bali proudly pointed to the flank of a large hill, planted with bananas, from which rose the neat houses of the town of Bali, like beehives in the distance.

Freshly strengthened by our breakfast, we walked ahead, and after an hour's march we entered the centre of Bali, surrounded by palisades, through its one gate. The town is built on both flanks of a hill on the very top of which is situated Garega's palace, which fronts a big marketplace. From here starts a broad main road which goes along the ridge of the range, on which are built, symmetrically facing each other, one hundred houses on each side. These are the residences of Garega's wives, followed by the houses of the king's sons, the elders, the dignitaries of the people, and so on.

Having reached the marketplace, we stopped in front of the king's palace and fired three times with 25 guns, a thundering salute. Then we wandered further through the town, surrounded by shouting and dancing people, who threw palm leaves on the ground, to Zintgraff's former station, where Garega intended to welcome us. This sensitive concession, to welcome us so to speak in our own home, had a particularly pleasant effect on us.

Arriving at the station, moreover, we were very surprised to see, rather than the heap of rubble which we had encountered at the other stations, a well-tended banana farm and five new huts!

Garega sat upon a throne of ivory; in front of him was spread a costly carpet. He sat like an iron statue, without even batting an eyelid, directing his glance fixedly and sternly upon us. At his left sat about 40 old men, the elders of his people, holding huge tobacco pipes between their knees. At his right small stools were placed for us. About 1000 warriors sat, in complete silence, in a wide circle on the ground, their guns and spears held upright between their knees.

Garega wore a reddish purple, mantle-like Hausa garment, the folds of which hid the shape of his body completely. His face

Figure 7.3: Mfon Galega of Bali-Nyonga – a photograph by Zintgraff reproduced in the *Koloniale Zeitschrift*, 1900.

showed strikingly few Negroid features. A pair of clever eyes sparkled beneath a noble forehead. On his head he wore a little raffia cap, a necklace of beads adorned his neck, and on his feet he wore sandals.

He got up and came towards us, shaking hands with each of us. Then he sat down again and, through his interpreter, bade us sit down on the stools to his right. After we had followed his bidding, the people clapped their hands three times with bowed heads to greet us. Thereupon, Garega took two kola nuts, chewed them into mush, spat this upon a banana leaf and sprinkled a few grains of kubeb pepper on it. Then all of us had to partake of this mush to prove that everyone would speak the truth, and that neither we nor he harboured any bad intentions against the other.[22]

Then he took a buffalo horn full of palm wine, sacrificed some drops which he spilled on the earth, took a sip himself and offered us the horn, while telling us the following through his interpreter:

'The king welcomes the White men in his country. The Whites are clever and powerful; therefore the king loves them, has ordered houses to be built for them, and he hopes that they grow fat during their stay with him, for the march seems to have weakened them.'

We answered him that his, Garega's, fame, and the knowledge of his wisdom and power, had reached us far over the sea in Germany, and that we had made up our minds to visit him and to discuss with

Figure 7.3: Bali-Nyonga seen from Baliburg – a photograph illustrating an article by Franz Hutter in the *Koloniale Zeitschrift*, 1900.

him things which were of importance for his country. We would accept his hospitality with pleasure, and hoped to grow fat again, to his full satisfaction.

Garega seemed to like our answer; we had flattered his vanity. Then the horn was filled yet again and twice while nobody uttered another word. The king broke kola nuts and distributed them among his old men. Thereupon he again recommended us to eat enough – we really must have looked very hungry at the time – and went back to his palace. The master of ceremonies now introduced us to the men of importance; at their head came M'Bo, the heir to the throne,[23] then the other sons of Garega and some of the councillors.

Then we repaired to the huts which were assigned to us, to give ourselves a rest.

In the gathering darkness, Garega's presents were brought to us: they consisted of four women, ten pigs, capons, cocoyams, maize flour, several calabashes of palm wine and, yes, potatoes. It was a big surprise to find the last, of exquisite quality too, up here. After further investigations we soon learned that they had been planted at the time of the Old Station and were now generally cultivated by the Bali.

The next morning I went for a walk through the village which I had glanced at only cursorily on my first march through. It was market day, and the Bali were trading in slaves,[24] ivory, knives, spears, and dogs. The dogs, above all, constituted an excellent item of trade. In the swampy jungle they do not do well; the Bakundu, however, relish them for their orgies and exchange them for sheep, goats and, above all, gunpowder, lead and salt which come up from the coast.

The most important article of trade in the protectorate is salt. Before the time of the Europeans, salt was produced at the coast by evaporation and brought up to the interior. Nowadays the natives buy trade salt from the factories at the coast, mix it with ashes and earth and trade this unrecognizable, grey-black mass in little conical packets. Far further away in the interior the tribes obtain their salt from the natron lakes of the Sahara or extract it from saline plants.[25]

As I have mentioned, a broad road runs from Garega's palace along the top of the hill; alongside it lay some 200 houses occupied by his wives. From this narrow road lanes branch downhill on both sides and lead to numerous large enclosed compounds. Each of these contains a dozen or more houses and thus constitutes a little community of its own. The houses are rectangular, built of clay, and surmounted by steep pyramidal roofs, which makes them look rather like beehives rising out of the surrounding banana groves.

The owners of the particular compounds greeted us in friendly fashion and invited us to enter their palm wine halls where we drank palm wine with them amongst the calabashes stored there. Hospitality is one of the main virtues of the Bali. And it so happened that here, among these wide-awake people, I often lost any sense of being among wild Black folk, deep in the interior of Africa.

The nature of the country must surely reflect that of the inhabitants. These men were proud and brave with a free, open outlook, long-legged sons of the mountains, of the highlands, wiry and muscular in build. The flabby bodily forms of the dwellers in the steamy forests, who were spiritually as well as morally slackened and who had sunk to a lower level, are not to be found hereabouts.

The shape of the skulls of the Bali is characteristic. Soon after birth they put pressure upon the upper forehead in order to mould the back of the skull into an egg-shaped form. Incidentally, this habit is found among many of the border tribes of Adamawa.

The two front teeth of the men are chipped to a point, while those of the women are pulled out. Both sexes shave off the hair on their heads either completely or – and this is done mainly by women – they retain a caterpillar-like ridge of hair along the middle of the head, which looks rather like a little fur cap. In times of war the men grow a tuft of hair on the crown of their heads in order to enable the enemy, when he has cut off the head of a slain person, to carry the head comfortably and does not need to stick his spear through the ear or the mouth; it is seen as humiliating to be disfigured after death.

The women have pierced ears and lower lips, and wear shells in their ear-lobes. Their lower lips are transfixed by a small wooden plug which reaches from the nose down to the chin and presses the lips apart from above and below and makes them quite swollen.

The skin-colour of the Bali is blue-black. However, the king, as well as other dignitaries, rub their skins with red camwood which, in time, gives the body a beautiful claret-red hue.

All the women use this red camwood – the only luxury they know. They go completely naked apart from a cord tied around their waists, to which, on the occasion of big dances and festivities, the *guassi* [sic], a plait of palm bast, a hand's breadth wide, or a bunch of grasses, is attached at the front and back. This is the customary decoration of the Bali women and, simple as it is, it is only put on at big festivities.

The *guassi* is generally made of banana fibres and sweet-smelling herbs. The one at the back is preferably adorned with a sort of comb around which are tied red, white and black coloured grasses and which stands upright like the outspread tail of a peacock.[26]

The progress of culture, however, has begun to affect the fair ladies of Bali; and Garega observed with alarm that the women, seduced by our presence, had started to be discontented with their *guassi* as their sole decoration, and that they had become demanding enough as to want to wear beads as well. It was on the second day of my arrival that Garega poured out his heart to me about this. His two hundred wives had heard about our beads and would not leave him in peace any longer, begging for these decorations. I promised to help him and sent word to the king's women that, in exchange for their spears and *guassi,* they would obtain beads from me. The king's women, incidentally, are allowed to carry a particularly beautiful type of spear. I had no idea of what I would provoke with this! During the heat of midday I had gone off to rest in my hut upon some hides, and had dozed off into a sweet sleep, when some terrifying female shrieks woke me up. Curious to find out what was up, I opened the sliding door of the hut and they all stormed and pushed in, so that the clay walls cracked and the hut shook. In a flash the entire room was packed tight with Black ladies, who, pressing me into a corner, put their sweet-smelling *guassi* under my nose. Meanwhile more and more pushed their way in. Fortunately my hut had a second door, which I tore open in wild desperation, hoping to be able to escape into the open air. With the help of my servants, I managed, with great difficulty, to get the Amazons out of my house. I positioned two Vai boys as guards in front of the entrances and then allowed the excited ladies to enter one after another in good order. Thus, I managed to buy their spears and *guassi*.

The iron for the spears as well as for all other iron tools made in Bali, is obtained from nearby Bafrum [Bafreng] and derived from surface ores.[27]

On that very day we had to arrange a particularly interesting, but also a rather worrying occasion, the presentation of the gifts we had brought for Garega – worrying since I did not know whether they had suffered damage or loss during the voyage.

The expectant king had already announced that he would visit us in the afternoon. To our relief the elegant and ornate throne which we had brought along soon stood, intact and correctly put together, in my hut; as soon as one sat down on it the prettiest melodies came forth (for example, '*Ach du lieber Augustin*' and other such). Then the Blacks enthusiastically polished up the cuirass and helmet, sword and gun, which we had taken along for their king.

Some of his counsellors had whispered to us that His Majesty especially loved the sugar of the White man and had been longing for it for years in vain. So we sacrificed some pounds out of our supplies. In addition, we were to present some cloth and several bunches of beads, the latter after I had got to know the hankering for them of the royal wives from my own experience and had felt sincere sympathy for the king.

Around four o'clock Garega arrived in grand processional style at the station. Leading the procession walked three slaves with fly-whisks, three with horse-tails, then six loaded with big bundles of spears, and following them a crowd of warriors wearing flowing Hausa robes. Then came the king himself together with some of his favourite wives who carried big calabashes of palm wine on their heads. I never saw the king without palm wine and wives in his company during my six days' stay in Bali. After the wives carrying the palm wine, came the dignitaries and senior sons of Garega.

The entire procession stopped in front of our hut. Garega entered and I indicated to him that he should sit on his new throne, which he carefully inspected from all sides. He was startled for a moment when the music started to play, but then seemed to take it for granted, and gave the impression that he knew very well that the chair was going to play music and that nobody could impress him at all.

Meanwhile the palm wine was warmed in a big earthenware bowl set on the burning logs in my hut and was offered to us by the women. Then I presented the king with the other gifts. When he saw the heavy broadsword his eyes shone with pleasure and he tried the steel with the air of a connoisseur. Altogether, he seemed to be very contented and stayed with us for several hours until late into the night, in the course of which about 50 litres of palm wine were consumed by him, his followers and ourselves.

His sympathetic character, his engaging friendliness and simple dignity made a tremendously pleasant impression on us. I detected only one weakness in the character of this African ruler and that was an indescribable vanity about his dignity, his power. He harped again and again on his importance and his strength.

Then we discussed with him the question which meant the most to us, the question of labour supply. At first he did not want to hear anything at all about it. He would need his people to wage wars, he could not send them away. I then tried to convince him how the power and prestige of his people would grow through work in lasting peace, how the prosperity of his country would increase, how he himself would have a higher income, and to put it at its crudest, how he would always be able to enjoy sugar and palm wine. It was the sugar that had the most effect and he began to consider how the proposal would be of advantage to himself and therefore as more acceptable, and promised to discuss everything with his elders.

Then he posed many questions about Germany. I showed him a portrait of the Kaiser in his uniform of the *Garde du Corps*, and he seemed to be delighted that he now owned the same warrior garment as the Kaiser.[28] Continuously looking at the picture, he then inquired why, if the Kaiser was a rich man, he did not wear any decoration, why he had not even a string of beads around his neck. This was a question which was for a moment very difficult for me to answer. Why? Because I could not tell him that beads were of no value or unfitted as a decoration for men, as I had just some minutes earlier put a fine string around his neck. So I told him that our Kaiser was a very good man, that he had just one wife, but that he loved her so much that he had given her all his beads as presents. At first this did not make any sense to him, but when I showed him a picture of the Kaiserin, in which she chanced to wear a string of beads, he was convinced and said: 'Your Kaiser has a beautiful wife!' Then we fell to talking about the question of women, and then about religion, and Garega said: 'The religion of the White man is good, because his God is more powerful than our gods, but he permits the White man only one wife, and this is not good. One wife is not enough for a man.' He further opined that it was wrong to pay so much attention to God, because this one was, as I myself had said, a good man who loved the human beings and would not harm them – but the bad God, now, the Devil, deserved to be worshipped, and one should be on good terms with him for otherwise he would bring disease and lost battles. And when we replied that the good God had power over the bad one, he said that therefore it was wiser to be on good terms with the bad one, because he was cheaper to win than the good God because he was subordinated to him.

Then he told us that his people, even under the reign of his father, had had a great religion, and how this religion, together with the homeland, was lost when they had to flee to the west from the Hausa. Now they did not really have a religion any more, and that he, like his forefathers, did not want to profess the religion of Islam.

In the 'forties of this century, the Bali had to give way before a religious movement which broke out at that time, the advance of

Islam against the heathen peoples of Central Africa, and had to withdraw further to the west, up to the very borders of the grassfields. Although they had fled from the Moslems, I found that the Bali were influenced by their religion, rather to their advantage.[29]

For us, these conversations were of the greatest interest even though we had them through two interpretations and they were therefore a little cumbrous.

I asked Garega whether he had ever heard of the missionaries, which he claimed he had not. After I had explained to him the fine profession of these men, he said: 'If your brothers were to come to me, I would believe their words, which are also your religion, and I would be baptized with all my people. They must be good people, these missionaries, and when they arrived I could make a strong medicine with them to reconcile the Devil.' Apparently Garega feared the Devil a great deal. This very Devil, just two months ago, killed his eldest son, and he would always bring him nothing but misfortunes.

The next day Garega had bidden all his people and the tribes subservient to him to come to Bali in order to show them to us, and us to them. At about 12 o'clock I was called to the palace, after Garega's master of ceremonies had been with us at 11 o'clock to beg us to come well dressed and not in such poor clothes, since the king would be wearing his cuirass and a splendid robe.[30] It was hard to decide what to do. We had nothing with us but our travelling outfits and even these were no longer wholly presentable. Finally I had a lucky thought. All of us were in the possession of a terry-cloth bathrobe which was provided with a broad, knitted red border and a hood and which had served us as blankets during the night. Additionally, we all had bedroom slippers of red leather.

We produced turbans out of highly coloured cloths, put on our bathrobes and slippers and, to enhance the festivities, we strapped on our cartridge belts and holsters. We must have looked incredible, but we gave rise to great admiration among the Black onlookers.

We, too, could display some pomp: first of all I made 25 of the strapping Vai boys march up in red jackets and blue caps, all supplied with an M71 rifle and three cartridges, from which they had had to take out the bullets with their teeth.

Then followed the women Garega had given us 'richly dressed in beads' with neatly made *guassi* tied around their waists, or rather fixed front and back at their waist-strings. I had asked Garega immediately after he had presented us with the women – incidentally a useless present for all of us[31] – whether he would allow me to dress these women in cloth. Nearly falling to his knees and with visible signs of alarm, he begged me not to do this, because from that very instant all his 200 wives and his more than 100 daughters would also demand cloths and ruin him. In his opinion, dresses were not healthy for women and cloths existed only to adorn men at war dances and at the celebration of a victory.

Our Vai boys were followed by two boys with foghorns, who were instructed to blow them constantly; then came Zintgraff, Hoesch and myself, and finally the 25 Bali soldiers, drilled by us, in blue jackets and white caps, also supplied with M71 rifles and three bitten-off cartridges. After them came 'the masses'. I could not help being reminded of the carnival parades of my home town of Cologne.

The costumes in which we paraded through the town, where, on this day, at least 10,000 people had assembled, were admired and gazed at by everybody, and greeted by the wild warrior cries of the men standing along the roadside and by the monotonous squeals of the women.

Now the Bali stormed into the marketplace, their heads adorned with huge feather crowns, their bodies painted, with colourful cloths round their waists, holding three spears in the left hand and in the right hand a Dane gun hung with war trophies and decorated with brass nails. A sort of festal joy beamed from their eyes.

Arriving at Garega's house, our entire procession marched into his courtyard. The ruler was seated in the palm wine house in costly robes awaiting us, and was visibly delighted to see us parading in such a fine and solemn manner.

We thanked him for this splendid celebration and the impressive war play which he had allowed us to see, to which he did not say anything in reply but emptied his horn smilingly, ordered it to be refilled and gave it, as a sign of his particularly gracious mood, to the four women he had presented to us. They beamed with joy at this benevolence on the part of their ruler.

Then Garega clapped his hands twice and three of his court jesters jumped out of a side room, entertaining us with a display of their tricksters' arts, while the old man withdrew.

Garega's court jesters are an especially interesting phenomenon and resemble the idiot scouts of the Negus of Abyssinia. They play the fool professionally and are considered as idiots by the people. They appear at all great festivities, play the fool, throw earth at themselves, clean the people with fly-whisks, or beat them with their batons, utter inarticulate sounds, hug the women, stare foolishly into space, jump up and down and turn a cartwheel; in short, they act the fool.

Garega owns over thirty of these men, who are recruited from amongst the most intelligent people of the country and who wander freely and unhindered along the borders and inside the country of neighbouring peoples, observing everything with a sharp eye. For as fools they enjoy a fool's licence. Nobody dares to harm them or prevent them from entering a neighbouring country. But they keenly observe everything that is going on in the neighbouring tribes and report it to the king. In a way they take the place, if I may allow myself an analogy, of the office of envoy or military attaché in our modern states.[32]

Soon Garega reappeared, wearing the cuirass and helmet and a creased red gown. The shining steel suited this hardy tempest-proof old man extremely well – he must have been nearing his seventies.

Pointing to his neck which lacked the beads which we had given him as a present, he said: 'I want to be as good as your Kaiser, my protector; I have given the beads to my wife as a present.' And in proof of this, his favourite wife, Babula by name, had to enter. She was embarrassed and happy at the same time and played with the beads around her neck. Thereupon he drew his broadsword and led us out to the market place.

What a surprisingly colourful, exotic scene presented itself to our eyes! The extensive marketplace was surrounded by thousands of fantastically garbed people forming a large semicircle starting from the right and left of the ruler's house. When the appearance of the old man was noticed, there was a dead silence. Garega and we sat down in front of his compound, he on the throne, we on stools, and then hundreds of children, carrying palm leaves in their hands, closed the semicircle behind us.

All the people received us, as always, in a squatting position and clapped their hands three times to welcome us, whereupon the ruler's favourite hornblowers started to sound their large ivory trumpets. Then, all of a sudden, the court master of ceremonies, feather-crowned, leapt half-way across the open space, shaking his spears threateningly in all directions. The semicircle of men surrounding him responded with wild flute-play, drumming and with battle cries, all at once. He was followed by all the king's councillors and the elders of the realm, who had been squatting on the ground to our right: they exactly repeated the movements of the dancer in the lead.

Now Garega stood up, the sword in his right hand, his horse-tail fly-whisk in his left; behind him were his favourite slaves. Like a roaring thunderstorm, all his warriors flung themselves in his direction giving loud battle cries and then halted, tightly packed, just in front of him, holding their spears threateningly in the throwing position. The slaves[33] jumped in front of their ruler, forming a living shield. Beating their chests, they boasted of their loyalty to Garega, and of being ready to die for him. The warriors replied threateningly with counter-cries, all in rhythm. An exciting picture.

As the old man brandished the broadsword there were thundering cries all around us and all the drums roared; the lines of warriors came forward and went back three times, as if attacking the ruler. Then from higher ground on the right, like a sudden whirlwind, M'Bo, the successor to the throne, stormed down with four hundred picked warriors to protect his father. Many shots were fired.

At the left hand of the ruler, protecting him, now stood his powerful son, surrounded by his warriors. They pushed forward like a wedge into the thousand-fold throng. There was a constant movement backwards and forwards, a surging to and fro, which was executed in strict rhythm to the music. Then the signal to stop was given, and Garega started to criticize the war game.

After the king had sat down again, the warriors formed two lines, facing each other. The warriors stood as if preparing to spring forward, the left knee bent, the right leg tensely stretched, ready for single combat. Then some rushed up to the front line, to show off their agility and strength. More shots were fired, some pretended to be hit, while others leapt towards them, drew their knives and imitated the movement of decapitation. Then other warriors leapt out of the lines, and this went on until all of them had exhibited their agility.

The old king kept his eyes firmly fixed on his smart warriors: he knew every single one of them well, and the interpreter often urged us, on behalf of the ruler, to watch them closely and to pay special attention to this or that movement of the men.

Now, after each of them had shown off his skills, both opposing lines ran towards each other, brandishing guns and swords, and almost crashing into one another, then turned to the left and right, and in a few seconds stood in warrior-like postures and with marvellous precision in neat order before the ruler.[34]

Garega once more criticized the performance, dispensing praise and blame. And then I asked him to allow me to show him my soldiers. He agreed, with eager expectation.

The first march-past, with the Bali, I led myself; the second was led by my companion Hoesch. The parade-marching was perfectly executed. In our bathrobes, however, we must have looked quite peculiar to European eyes. The firm ground of the marketplace rang under the steady steps of the barefoot Blacks, and Garega was highly contented when we made the formations march towards each other and fired a salute. Thousands of delighted voices answered the roar and bang of our guns. Thereupon we swarmed out in skirmishing order and opened a lively but scattered fire. To conclude the parade I made all fifty men file past Garega in a slow march.

After the lively warlike performance I had just watched, this exercise of ours seemed to me unusually sober and civilized. But the glee of the old man was boundless; he continuously slapped his knees as a sign of his approval.

Garega now gave the order for the dance to start.

The ivory horns sounded yet again. The heir to the throne himself ran with a large drum to the centre of the marketplace to play for the people to dance to, and the old king opened the round dance with elastic steps. Then we too were irresistibly swept up in the general joy and enthusiasm. Guns were going off everywhere so we, too, took out our revolvers, and in the midst of the rejoicing people, jumping up and down behind their ruler, fired shots into the air. Then, suddenly, peculiar melodies could be heard. Singing strange songs, out came Garega's wives from the palace with rattling, hollow, iron ankle-rings filled with little balls, and approaching in single file, formed a large circle around the king and ourselves. It was a sort of Polonaise with many undulations, performed very gracefully by men and women alike. Finally the children followed on, and it was delightful to see how skillfully toddlers barely four years old comported themselves, keeping time as dead seriously as their elders in the processional dance. They bore palm branches or bunches of green leaves in their little hands which they waved gracefully.

The people amused themselves by dancing and consuming vast amounts of palm wine, which the old ruler had ordered to be brought up from the surrounding neighborhood to celebrate the feast, and meanwhile Garega invited us and his dignitaries into his house for a delicious meal.

The custom of the Blacks to pick food from the dish with their fingers is far less crude than might be supposed, being marked by the absence of any greed or haste. They have a peculiar knack of supping up even liquid food without a spoon just by crooking the forefinger.

Shortly before nightfall we said good-bye to the old man, while the people outside continued dancing indefatigably to the sound of the drums and by the light of the leaping flames of bonfires until far into the night. The next morning provided me with examples of the Negroes' keenness to visit people and of their easily-aroused covetousness. The day was barely dawning when, with the rising sun, many of Garega's councillors turned up to beg for presents from us. The knowledge of the richness of our presents had spread like wildfire in the town, and, although it raised our prestige, it also aroused the greed of the dignitaries of the court in a marked way.

We did not get involved with them, and as soon as their molestations became unbearable we addressed ourselves directly to Garega with a request to restore order. Whew! The old man was really furious! He took a bunch of spears from the slave next to him and hit out at the dignitaries who squatted there dumb and meek, with bent backs, until the spear-shaft splintered to bits. Then taking a deep breath after this unwonted exercise, he said: 'You White men, you can leave your treasures everywhere around in Bali without supervision. Anybody who takes something away or starts to beg again will be sentenced to death.' This was a 'king's word' – nobody asked us ever again for presents.

That afternoon we negotiated again with Garega on the question of labour and came to the following result:

He, the ruler, would allow several hundred people a year to go to the coast: they would have to pay him a head tax when they left as well as when they returned. In return, each Bali would have the right to take to the coast with him five men free of tax from the tribes which were subject to the Bali. But he would be bound to pay the agreed bridge toll at all hanging bridges along the way for these men.[35]

Furthermore, the ruler made it a condition that the Whites would see to it that his people would not be molested or attacked by other tribes. Not long ago, eight Bali were caught and sold into slavery in the forest of Sabi by people of the Babe tribe, who were alleged to live in a town to the west of Sabi. Garega's messengers had returned without having achieved anything, and the old man had already secretly prepared for war. I promised him then that if he would accept my wishes I would totally satisfy his demands and send back the captives.

The course of this affair will be described later.

We had already spent five days in Bali, and now it was time to think about the march back. The swollen rivers and torrents, the soggy and swampy ground heralding the beginning of the rainy season, had already provoked great troubles on our way up, and could become virtually fatal on our way back, if we postponed our return for too long.

It had now started to rain for eight hours a day. And we did not have any tents either, to provide for some commonplace mishap which might hold us up by day and force us to spend the night in the open.

I explained to Garega that we had to leave the next day. Then the old man looked sad and said that as his guests we had had all we needed and that we should stay, and that he would send us more women, pigs, dogs, fowls and palm wine. We had not yet recovered and were still very thin.

But I and my companions kept to our decisions; only Dr Zintgraff was to stay behind in order to lead the first workers himself to the coast when the rainy season had abated, and meanwhile he was to start up kola plantations in Bali.

In the interest of the colony and the plantations, Zintgraff undertook the difficult task of holding out all by himself up there for four months in order to bring down the first workers to the coast as we had planned. In doing this he again rendered outstanding services to Cameroon. We, however, quickly made preparations for our journey and the next morning saw us standing, ready to go, in front of Garega's house to say good-bye to the ruler. Thousands of people stood tightly packed on the road and the marketplace; and everywhere along our way they strewed palm leaves and aromatic herbs. Garega received us in the palm wine house and presented some ivories to us as a farewell gift. Then we discussed and swore to everything again, drinking palm wine and chewing kola nuts as usual, and finally Garega asked once more for a missionary, to produce with him a medicine strong enough to kill the Devil.

We promised to do whatever we could.[36]

And then came the really moving farewell scene. The old man said: 'You, White men, I probably will not see you again, but you will fortunately return to your homeland, because your hearts are brave and your foot is strong. I will bless you, so that luck and health go with you. And when you finally return to Germany, tell your Kaiser and all big men about me, a mighty sovereign, who will always remain faithful to the Whites.'

Thereupon he escorted us to the front of his house, took a wooden bowl full of water, pouring it on our heads saying: 'See this clear water: may your eyes see as clearly so that you do not suffer any damage.' And sprinkling our backs, he said: 'As I wash away the dust from your back, I also wash illness and death out of your body.' And then sprinkling our chests: "Because through the water everything grows on earth, so let you prosper and remain powerful and kill all your enemies." The rest of the water he poured partly on our feet, partly on the threshold of his doorway, crying with loud voice: 'As the water puts out the fire and eases the pains of wounds, let it protect you from all pains and strengthen your feet, that they may carry you to your faraway homeland, and let it be that your feet will never have brought or ever bring any misfortune either over my threshold or over that of one of your friends.' Then he embraced us all, and put in my hand an ornate staff and said: 'If you don't like to stay anywhere else in the whole world any longer, come back to us bearing this staff, and if I am not alive any more my people will recognize you by it and with the help of this staff they will take good care of you.'

Wild applause arose from the people and the Bali who had accompanied me performed a warrior's dance once more in Garega's presence. I then gave the sign to fire three thundering salutes, and since the sun peeped curiously through the clouds at this instant we could even take a photograph of the event.

Then another hug, a last sparkling of the old man's fiery, clever eyes, a last waving of the clenched fist, the sign of welcome and goodbye in Bali, and we were off again, confidently awaiting new dangers and new hardships.

Six dignitaries escorted us to the border of the kingdom. Then, we were on our own again, and off we went in speedy marches down the airy heights and into the fever-impregnated jungles towards Tayo, to bring the enemies of the Bali to judgment. As I mentioned above, I had promised Garega to call in there because of the previous abduction, in order to prevent him from undertaking the bloody war he had already prepared.

We followed the old route until we reached the forest of Sabi. Then at approximately half the distance between Sabi and Miyimbi, we turned west in the direction of the German-British border, into an area never before penetrated by Europeans. People from Sabi acted as our guides.

The reason why the very rich area of Tayo had so far been shrouded in darkness was the fierce character and the marauding spirit of its inhabitants, who live cut off in their crags. Their reputation for cruelty so frightens the Blacks, travelling on the trade routes, that they only dare to cross the forest of Sabi in large units and armed to the teeth, and they march for up to eleven hours a day in order not to have to spend the night there and to avoid contact with the land and people of Tayo.

At first we had to overcome a deep, torrential river. To cross rivers, across which, as was the case here, there are no hanging bridges, can be very difficult. First of all the strongest and tallest men have to get to the other bank of the river. They wade in pairs, their arms held high, taking a load to the opposite side. Often the current draws them right under water, and then the business becomes quite dangerous. After all loads have been taken across, and often the men have to cross the river three to four times in both directions, the smaller and weaker carriers are taken over at a place further up-river, at which the river makes a bend. Here all jump into the water and drift across the river on the current which, at such places, carries them to the other bank. Swimming is out of the question. On the other bank the men who have already crossed pull the floating men out of the water. It will be clear that this mode of crossing rivers sometimes leads to losses, but there are no other means. One is thankful every time to have crossed without fatalities.

In this forest we found an undreamed-of abundance of game. We saw elephants standing in the undergrowth, watching a silent caravan with their little eyes. I had prohibited all shooting in order not to advertise our caravan; and I had much difficulty in fighting down my passion for hunting.

Gradually the path rose, and became steeper and steeper. We were again at a height of 700 m, marching in single file through thick forest. Suddenly a shot was heard, then four, then five. I gave the order to halt and to release the safety catches, believing that our rearguard was involved in a fight. Then a huge elephant crashed, banged and thundered through the undergrowth, knocking down the intervening trees to its right and left with an unbelievable speed. My people had shot at it despite my prohibition, but had not wounded it mortally. In response to my stern inquiry they asserted they had shot in self-defence. The elephant, which had a calf, had charged them furiously.

Since the leaders assured me that we were still at quite some distance from Tayo, I hoped, despite the shooting, to be able to take them by surprise, although I retained some doubt about that given the excellent hearing of the Black folk. Soon we approached large plantations of groundnuts and cocoyams. The soil here gave evidence of great fertility. We might have walked through the plantations for some twenty minutes when from all the surrounding heights battle cries rang out and black figures could be clearly seen running to and fro.

I hurried on, giving my gun to my servant with only my revolver strapped on, to an open space which was visible from all directions, raising both arms and so showing signs of peaceful intentions. But in vain; none of them dared to come down. Then I stopped the caravan and went ahead with a few people, guns hanging over our shoulders. Soon we had mounted a height, but we could not see any of the natives; only the sound of their monotonous warning cry was heard.

In the distance we saw on various hills, in full sunshine, some pretty villages, from which women and children, loaded with their poor belongings, were fleeing in a hurry into the bush. Giving the caravan a sign to follow, I rushed ahead with my few attendants.

Arriving in one of these villages, I found it, as well as the others, deserted. My people, however, succeeded in catching one man hiding in the bush, who was brought to me, trembling in every limb. My interpreter found it hard to convince him that he would not suffer any punishment, and that we had only come to demand compensation for the injustice done to Garega, and with the aim of making a permanent peace between Bali and the people of Tayo. I released and sent the man to the chiefs of Tayo with this message, adding that we would stay until everything was settled; the natives should return, and we would neither plunder nor do any other damage to them.

In the meantime my people started to establish themselves cosily in the various villages, while I took special care to ensure that nothing was damaged or ruined. My instructions to the people of Tayo very soon brought about the desired result, since, one after another, they came down to us from the bush, particularly when they saw that my people had completely refrained from damaging their possessions.

Soon the two chiefs of Tayo,[37] called Tanyan and Atyan, appeared, surrounded by a body of about 30 warriors armed with guns; the hammers of the guns were cocked. We met on the open space. I walked up to them without any weapon: on one side were the people of Tayo in a semicircle, on the other side 25 of my Bali men, also with loaded guns. At first, the chiefs tried to deny everything. Then, after the witnesses had spoken, they realized that this attempt was futile, and they frankly admitted they had caught seven Bali and declared themselves prepared to restore these people to Garega in terms of ivory, for they had already sold them into British territory. Then I made one of Garega's sons, who had come with us, estimate the value of seven people in ivory, and soon we saw five enormous tusks, each taller than a man, propped up against each other like guns.

The people of Tayo and Bali now solemnly slaughtered two goats, the blood of which was collected in a large vessel. Then we all washed our hands and faces with this blood and poured it upon the earth, while both parties performed their war songs, dancing with abandon and brandishing their weapons. Thereupon everybody shook hands, chewing and consuming kola nuts. Their former owners gazed sadly upon the beautiful ivories, now being carried away by the Bali – and peace and friendship were promised for ever!

Now the women came out of hiding too, laughing with embarrassment, and getting used to their new lodgers.

At 9 o'clock, as a precaution, I evacuated our people from the entire village and set up double outposts, supervising them myself for the whole night. On the part of the natives nothing happened – they had obviously accepted their fate, and on the next morning they seemed happy to escort us for a while. The Bali, too, were now firmly convinced that since blood friendship was established, they did not have anything to fear from this tribe. Unfortunately they were to be deceived in this optimistic expectation.

As predicted, on our march back, the torrents and rivers presented us with big problems. Not only were our garments tattered and rotten because of the dampness, not only had our shoes suffered a great deal, but a great part of our food supplies had gone bad or were lost at river crossings. The worst of all our deprivations was that for ten whole days there had not been a single atom of salt left to eat. Moreover, we were devoid of all fats to roast with and of all flour for breadmaking.

After great hardships we arrived, very exhausted, at Mundame; but here our hardships had not yet come to an end. It was impossible to find enough food for the great number of our people, so I was obliged to shoot some of the oxen belonging to the factory, which were eaten up to the last morsel in three hours – a costly pleasure since such cattle are very rare in Cameroon. The next day we wobbled about in native canoes along the flooded Mungo: I had obtained them with the greatest trouble at a high cost and now had the sure feeling that we had left all dangers behind. Around noon, however, a cloudburst came down, and the river rose enormously within a few hours, by some 4 meters. We tried to steer the boats to the bank, but in vain; the power of the immense current swept us along. The light dugout canoes streaked through flooded plantations, thick jungles, over the huts of natives. On one side of us giant trees were hurtled away; here an elephant wrestled with the waves, there fowls and goats. Everything was swept down to the sea.

After a 24-hour ride, ten hours of which we spent by night at sea, during which one man of my crew went off his head, the waves finally brought us close to Bimbia, to a point from which we were able to row to the coast at around noon. The other canoes also arrived, save one, unfortunately lost, together with all hands.

From Bimbia we moved to the plantation of N'Bamba, where we were received by Herr Rehbein most amicably. We slowly recovered and after a two days' rest we were able to walk to Victoria, and were received, after our absence of two months, with cheers by the Whites who had settled there.[38]

Thus, the purpose of our expedition was fulfilled: the route to the hinterland had been made more accessible and secure. With Garega we had signed binding contracts, which guaranteed the provision of workers for the coastal plantation companies. And the nicest thing about it was that all this was accomplished in a peaceful way; we did not once have to use our weapons, nor punish any natives. Though, as I have indicated, a few months later the people of Tayo did not keep their word. They believed that I – I had left Cameroon in the meantime – was transformed by their gods into a bird, which they interpreted as a sign that they were released from their oath.[39] So they again started to capture Bali and killed a number of them. At the instigation of Dr Zintgraff, the German government had taken cognizance of the affair and was preparing a punitive expedition to Tayo.

Insofar as the personal experiences I gained on our march are concerned, I will add some words to this account.

We should not deceive ourselves into thinking that the hinterland of Cameroon is poor and likely to become poorer and poorer, even without being helped to think so by our competing neighbours, the English and the French, who have, for decades, been in possession of the easily navigable waterways which lead from our hinterland into their territory.

We have come too late to Africa to establish colonies which rely exclusively on the profits of trade. The few places in Africa made for such purposes were occupied by other nations long before we even dared to dream of colonies.

Trade in Cameroon has to be created by agrarian and industrial installations, by German diligence and German energy, and this can only be accomplished at present at the coast by establishing plantations on a large scale.

Let me also warn readers not to entertain the notion that we possess vast territories within our colony into which we could steer German mass emigration. For such only German South-West Africa is suitable, although its value for this purpose is not yet sufficiently known to assess how many German farmers could establish a secure home there. The Pacific territories, East Africa, Togo and Cameroon will never be able to accommodate German mass emigration. It might be possible that here and there, in the highlands of this colony, some German colonist villages could be created in the distant future, though, in my opinion, they would never play a noteworthy role insofar as the question of German emigration is concerned. The dream of the geographers and early colonial enthusiasts of the acquisition of colonies of settlement for Germany has only come about to a limited extent. The question of a beneficial and practical emigration, which is so important for our Fatherland has still, despite the colonies, to be considered as unsolved and open. Let us hope that with respect to this, too, our national endeavours will one day be crowned with success.[40]

Notes

1. The name Garega is more usually transcribed as Galega by the Bali-Nyonga, who have attracted a considerable literature. An extensive bibliography is attached to Richard Fardon's *Raiders and Refugees* (1988) of both published and unpublished works pertaining to the origins, different trajectories, and ethnic composition of the Chamba-led raids and dispersals, and their ultimate self-definition. Among the Grassfields Bali, the Bali-Nyonga have produced their own literature and some of it, for example V. Titanji *et alii* (1988), N.B. Nyamdi (1988) and items produced by the Bali Historical Society, are discussed by Fardon (1996) in *African Crossroads*, 17–44. To these may be added a University of Pennsylvania thesis by S.W. Russell (1980), a compilation in English and Munggaka edited by J. Stöckle (1994) and continuing work by Eldridge Mohammadou on the 'Baare-Chamba' movements, discussed by Zeitlyn and Chilver in *JASO* 26(1), 1995. See also in this connection the contribution of C.-H.Pradelles de Latour in the same issue on the Pere (Peli in Bali-Nyonga) component and a section in Q. Gausset's Brussels thesis (1997) on the Kwanja and Wawa. A Boston University thesis by Sarah Richards, on women's attitudes to fertility and infertility, is completed. Several Master's dissertations by Bali graduate students are listed by Yaounde University. German published references, including interesting material in Basel Mission journals, are listed in Max Dippold's bibliography of German sources for Cameroon, 1971; see also Jenkins (1996) and Merz (1997). To archival sources in Berlin,

Yaounde and Basel may be added a copy of an administrative file for Bamenda Station, 1908–1913, with an English translation, in Rhodes House Library, Oxford, MSS Africana, also in ANY FA/110. This deals largely with the *Balifrage,* the Bali question; see Part III, 4 and 5.

2. Von Puttkamer, who visited Sao Thomé when he was Acting Governor in April 1895, gives a rosy view of labour conditions and of the role of Labour Inspectors; cf. *Gouverneursjahre* (1912) 47–48. He writes: 'Friend Spengler strengthened my resolve and declared himself ready to acquire land in the mountains and devote some of his time to the German colony. Consequently I then so informed the *Kolonial-Abteilung* from Monte Café and asked their permission to sell Spengler a larger land-complex, some 10,000 hectares, on a suitable part of the Cameroon Mountain for the establishment of a cocoa plantation. This permission was given.' This land, together with that acquired by Zintgraff at Kakaohafen, Sholto Douglas at Boana and von Soden in the Buea area formed the initial holding of the WAPV (Puttkamer, 1912, 102). While still Imperial Commissioner for Togo, von Puttkamer had advanced the view that Africans were 'incapable' of developing a cash-crop economy and later envisioned the system of virtual serfage for the Bakweri which he says Esser approved (*ibid*, 104–5). His proposals were somewhat mitigated as a consequence of what he describes as the 'unbelievable difficulties' put in his way by the Basel Mission and its allies in Germany who included the influential Hanseatic merchant J.R. Vietor, who shared their vision of Christian communities of free farmers. The Basel Mission came into direct collision with the WAPV in 1898 in the Buea area where it had planned to establish not only a missionary station and seminary, but a model Christian village, with full supporting services. See Rudin (1938), 365 ff., Halldén (1968), 92–139, and van Slageren (1972), 68 ff. for some discussion.

The Pallotine Mission, which arrived in the plantation area in 1894 and settled at Bonjongo ('Engelberg') gave von Puttkamer little trouble by comparison. It established good relations with the WAPV, which endowed a church, and concentrated its efforts on schooling, health care and practical training. See Wirz (1972), 204, 208, Hausen (1970), 237 for Esser's gifts, Skolaster (1924) and Wright (1958), 36–37.

3. A clear outline of the situation is given by W. G. Clarence-Smith in the *Journal of Southern African Studies.* 2 (2), 1976, 214–233. The agitation of British interests against the abuses of forced labour is discussed by James Duffy (1967), *A Question of Slavery.* For a sympathetic account of Portuguese attitudes, quoting Portuguese publications, see F.C.C. Egerton (1957), *Angola in Perspective.* A darker picture, based on local archives, is presented in David Birmingham's case-study of coffee-planters in the Cazengo district: see his collection of reprinted papers (1999). For overviews see Clarence-Smith (1985), *The Third Portuguese Empire,* and Eyzaguirre (1988).

4. Neither Esser nor Zintgraff (1895) mention the damaging report of conditions on the Bali road, published in the *DKZ* in 1893 (Nos. 1 and 5) by G.Böckner, who had been a coffee planter in Mozambique and was employed by the Botanic Garden. He was temporarily attached to Zintgraff's enlarged expedition. With Sergeant Knetschke he was charged with conveying ammunition and mortars to the beleaguered party at Baliburg, with an unsuitable carrier force of ailing Dahomeyans. For a brief account of the situation he described, see Chilver (1967a), 489–490. Villagers had retired from the unruly caravan route and refused to provide carriers or food, so messenger and carrier parties increasingly looted, while others mutinied. Böckner put the mutiny down to 'bad handling, reduction in wages and continuous beatings' (1893, 36). Zintgraff's scornful references to him are to be found in a letter to the German Chancellor, von Caprivi, of April 2, 1892 (ANY, FA 1/84, 213–216). See also a report by von Stetten (*DKB* 4, 1893, 34 in Part III) which mentions the withdrawal of villages from the route and the high-handed behaviour and looting by Bali trading parties plying the route, and the account given by Mission Inspector Lutz of the trek to Bali in 1903 in his *Im Hinterland von Kamerun* (1907), 8–10. The depredations of famished carriers and messengers was disruptive enough but even worse were those of unsupervised soldiers accompanying expeditions in later years to judge, for example, by a report from the GNK representative Diehl of 1903 and others from the Basel Mission giving examples from 1905, for which see Part III.

5. For a different and later viewpoint see a report by the naval captain Schwartzkopf (*DKZ*, N.F. XI, 1898, 43). This stresses the urgent need for a Station in the 'Bali lands' as soon as the Protectorate forces could be strengthened. He urges that this should be done before the Bali were tempted to embark on adventures nearer the coast 'which would have to be punished'. The only means of providing peaceful conditions for trade in the hinterland, he opined, was to establish a military presence strong enough to overawe the hinterland dwellers. This could not be achieved by 'peaceful little expeditions from a forward base'.

Esser's earlier attitude reflects that of the Hamburg interests represented on the nominated Colonial Council and of Kayser, then director of the *Kolonialabteilung*, who, in 1894, had advocated 'peaceful penetration' for scientific and commercial ends, now that the formal boundaries of the new colony had been established (*DKB* 5, 1894, 569–570).

6. Mundame, at the end of the navigable stretch of the Mungo, was now the site of a Jantzen and Thormählen factory and was described by Zintgraff in 1886 as a sort of port or beach for the Balung (Balong) who were middlemen between the Duala and the 'bushmen'; it was then a small village with only two slave settlements. Until a steamboat capable of carrying 25 tons was introduced in the 1890s the river-trade was carried on in surf-boats manned by 8 Kru-boys (Böckner, 1893). Zintgraff's

resited Barombi Station, where he had planted a vegetable garden and a field of Liberian hill-rice, had become an agricultural research station. For the latter, see L. Conradt in *Beiträge zur Kolonialpolitik* 1, 1899–1900, 322–346.

7. Here Esser appears to be repeating a question already put by Zintgraff in Ikiliwindi (or Kiliwindi, a Bafo village according to Conrau and later travellers) when he arrived there in February 1888. An attack on Schwarz's expedition was strongly denied on this occasion by its Chief, called Buamboke by Zintgraff (and succeeded by 'Porro' by January 1889). Zintgraff recounts that he was later told by Dr. Bernhard Schwarz's Swedish companion Knutson that the Ikiliwindi were engaged on a large-scale hunt for antelopes, which had at first been taken for a hostile demonstration. Knutson had soon been persuaded of the real circumstances. Dr. Schwarz was not. (Zintgraff, *Nord-Kamerun*, 75–79). Schwarz's own account differs; he attributes his final decision to turn back to a truculent message received through a Bakundu slave he had sent out to make inquiries (Schwarz, 1886, 323–327). Knutson's unpublished memoirs are still fuller. He had promised to assist the expedition as far as Barombi ba Mbu. Despite his advice, Dr. Schwarz had adopted an overweening approach, boasting of thousands of German guns and soldiers when at Mambanda and Ikiliwindi. This provoked a hostile but harmless demonstration by visiting 'Bafarami' (probably Bakossi) which rattled Schwarz and he was then badly frightened by the combined village antelope hunt his party met, to the amusement of his carriers. The Chief of Ikiliwindi had indeed warned Knutson that Dr. Schwarz, who had ostensibly come 'for the purpose of bringing German soldiers into the country', would be opposed if he went further inland. Zintgraff remarks that he felt obliged to protect his kindly host from the bad reputation given him by Dr. Schwarz. Knutson also describes him as 'a kind and nice old man'.

8. Zintgraff's interpreter, in 1888, described the place vividly: 'This place elephant use for dance and marry himself' (*Nord-Kamerun*, 83). Zintgraff saw well-carved 'idols' at Baduma where he hoped, without success, to acquire a man-high one. 'Ebulu' was reached next day *(ibid.)*.

9. See Plate 12, and Appendix II.
   The term 'fetish' at the time was applied to sculptures, masks, and other objects associated with secret societies as well as to the societies themselves (*Fetischbünde*). The standard work of reference was then Adolf Bastian's *Der Fetisch an der Küste Guinea's*, 1884, Berlin. For a history of the term see J. F. Thiel et al. (1986), *Was sind Fetische?*, Frankfurt, 15–31.

10. For an extended description of Bakundu (so-called) house-styles and village lay-outs, see Hutter's descriptions illustrated by Zintgraff's photographs (*Nord-Hinterland* 1902, 270–273).

11. The widespread belief in the cannibalism of strangers, evident in Cameroon folk-tales, was encountered by the Swedish hunter-traders as Valdau relates of a trading expedition they mounted in 1885. When he tried to recruit carriers for a journey to the Barombi lake and other places behind Mount Cameroon, earlier enthusiasm in his village for the expedition evaporated. 'It had come to their knowledge that the inhabitants of that land were frightful cannibals for whom any stranger was a welcome addition to the stewpot'. (*D. Geogr. Blätter*, 1886, 32). Of the Bakundu – the designation is loosely used by him – Zintgraff, too, thought 'that the suspicion was to some degree grounded', though praising their kindliness: see *Nord-Kamerun*, 83–87. Hutter (*Nord-Hinterland*, 357–358) is more cautious. Esser, as we see, in common with others, interprets the presence of trophy skulls and reports of the funerary human sacrifice of slaves as evidence for a festal culinary cannibalism, associated with a secret society. The government doctor Plehn (1904, 724) quoting earlier sources, such as Lessner (1904) in *Globus* on the Ngolo, asserts that 'all' the forest groups were cannibals. A later ethnographic survey of the Bakundu by the Basel missionary Bufe refers to trophy skulls and the former sacrifice of slaves at their masters' deaths, but makes no mention of their being eaten (*Archiv f. Anthropologie*, 12 N.F., 1913, 228–239). See also Mansfeld (1908, 158) on head-hunting in the Cross River districts and also, 159 ff., for the possible administrative use of their secret societies. As to the clothing and other insignia denoting membership of Esser's imagined cannibalistic society, sumptuary regulations commonly indicated the grade of an association reached by a member in the Cross River area. Even today only members of the superior grade of the Leopard Society are allowed to wear hats and trousers at an association meeting (Röschenthaler, 1993). See also references in the unpublished notebooks of Father S.A Carney, Mill Hill Missionary, for the 1930s and 1940s.

12. For an account of German policy on internal slavery and slave-trading see Rudin (1938), 389–396. See also a collection of essays on slavery and slave-dealing in Cameroon in the nineteenth and early twentieth centuries, edited by B. Chem-Langhëë in *Paideuma* 41, 1995, 95–272.

13. Esser appears to contradict himself here, since the labour treaty later described made provision for the payment of bridge-tolls.

14. The 'Batom' villages, according to von Stetten (1893, 33) then stretched 'from Kombone to Mabesse': Conrau designates Batom or Dikumi as a Bafo village while Migeod (1925) describes 'Bolo', Kombone and Dikumi as Bafo. Another 'Batom' village mentioned by Zintgraff is Kokobuma. 'Batom' as a Bafo clan name reportedly derives from Ntom, a legendary founder. Were one to rely on Hutter's descriptions of house-styles, what Esser describes pertains rather to the next group, the 'Mabum or Babum from Mabesse to Nguti (Sokwe)': this term may refer to the Bebum of the Nguti area.

15. We remark here on the dearth of easily available contemporary ethnographic and sociological literature for the inland forest groups of the South-West Province by comparison with the coastal area and Grassfields. This is less true for the fields of art, linguistics and, latterly, the political economy of forest conservation. For linguistics see the bibliography of Cameroon languages issued by ORSTOM and ACCT (French technical aid) in Paris and edited by Barreteau, Ngantchui and Scruggs, 1993, the preliminary linguistic atlas edited by Dieu and Renaud, 1983, and the more complete album of 1991 compiled by Breton and Fohtung.

    We have a survey of Kumba by Lagerberg and Wilms in 1974. For the better covered Kosi (Bakossi) we can cite J. Ittmann in *Africa*, 26 (4), 1956 among his many other and later works, a thesis (1976), and socio-economic studies (1977, 1980) by Michael Levin and a detailed study by H. Balz (1984), Basel. For the Banyang we have M. Ruel's study *Leopards and Leaders*, of 1969, and his earlier and later papers; M. Niger-Thomas' paper on indigenous credit and savings societies in Mamfe in Ardener, S., and S Burman, eds., *Money-go-Rounds*, 1995; B Chem-Langhëë and E. S. D Fomin, Slavery and Slave Trade among the Banyang, in *Paideuma* 41, 1995, and Fomin and V. J. Ngoh, *Slave Settlements in Banyang Country*, 1998, Limbe, with a useful bibliography of local archival and unpublished dissertation materials. For the Ejagham we have a number of publications by Röschenthaler from 1993 to 1999 with more to come. It is hoped, before long to make Edwin Ardener's survey of British administrative reports on the northern borders of the present South-West Province available as a starting point for further work. Locally produced publications exist, to our knowledge, on the Bafo, Nguti, Kosi (Bakossi), Mbo, Ekot-Ngba and Bangwa borderlands, and there are sure to be more, available in local bookshops, which deserve attention. Publications may be expected from C. Ifeka on the Takamanda-Obudu borderlands. Meanwhile see *JASO* (28) 3, 1999 for the Boki. For the issues surrounding participatory forest conservation and regeneration and the self-identification of modern communities see Barrie Sharpe (1998) in *Africa* 68 (1) and the references he gives. For modern elite struggles and popular recourse to witchcraft idioms in a North Bakundu village see a penetrating paper by Dickson Eyoh (1998) in *Africa* 68 (3).

16. The limit of navigability with more powerful steam-launches later became Mamfe.

17. For the establishment of Tinto Station in May 1892 and its use as a base for Hutter's operations with his Bali troop in training, see *Nord-Hinterland*, 373–374.

18. Both Zintgraff (*Nord-Kamerun*, 75–76) and von Stetten (1893, 34) record the popularity of snuff-boxes as articles of trade and means of exchange. Zintgraff mentions local artifacts made of land-snail shells

with wooden lids and small imported tin boxes commonly embossed with the head of Queen Victoria. Schwarz (1886, 148) was surprised to see snuff-boxes worn as a single earring by Bakweri women. As to tobacco, we learn from Mansfeld (1908, 15) that it was grown in the Ejagham village of Tabo and locally traded, and from Hutter, *Nord-Hinterland*, 395, of its cultivation and preparation in the Grassfields, with small quantities grown in the Banyang and forest villages he traversed, *ibid,*283.

19. Von Stetten mentions the great popularity up to the Grassfields of large black and white beads and small red beads. For those popular in the Grassfields see Hutter (*Nord-Hinterland*) 93, Chilver (1961) in *Afrika and Übersee,* 45, 235–258, and Pierre Harter (1981) in *Arts d'Afrique noire,* 40, 6–22. Currency and other trade beads are dealt with by Warnier (1985), 87–90, 142–3.

20. Mansfeld (1908, 220–223) gives an account of the widespread belief in transformation of a partible self. See also his briefer note in *DKZ,* 1909, 218–219, Ruel (1970) and Röschenthaler (1993). For a recent survey offering many Cameroon examples see Geschiere (1995).

21. The Kola species commonly grown in the Grassfields and preferred by Hausa and Jukun traders was *Cola anomala*; see A. Russell (1958) in *Tropical Agriculture,* 32, 210–240. For early mentions of the trade see E.R. Flegel (1885), *Lose Blätter aus dem Tagebuche meiner Haussafreunde* and L.H. Moseley (1899) in *Proc. Royal Geog. Soc. 14* (6), 633. Moseley was an agent of the Royal Niger Company, who visited 'Bafum', which included Bum, Mme and Esu.

22. For an extended account of Grassfields pact-making ceremonies see Hutter (1899a) in *Globus,* 74, 1–4, and in *Nord-Hinterland,* 488–489.

23. Tita Mbo, commonly known as Tita Gwenjang, succeeded Galega in 1901 as Fonyonga (or Ganyonga) II: in Bali-Nyonga regnal names have come to alternate regularly, Galega being the son of Nyongpasi or Fonyonga I. Fonyonga II died at an advanced age in 1940 and was succeeded by Galega II. Chilver's informants believed that the premature death of Tita Nji avoided a serious succession dispute. In 1893, Conrau had suggested that a conflict over succession might break out at Galega's death and render the reoccupation of Baliburg Station inadvisable (*DKB,* 1894a, 190). Exactly who performed the duties of the 'master of ceremonies' on this occasion is uncertain. One Fonte, head of the smiths and carvers charged with making the *lela* emblems, is frequently mentioned by Zintgraff, and also by von Stetten (1893) as maintaining the liaison between the station and the palace. He was sent to meet Ramsay's expedition in 1900.

24. Esser seems to be mistaken in reporting that slaves were openly trafficked. According to elderly informants interviewed by Chilver in 1960

and 1963, this took place in greater privacy; see also Hutter (1902) *Nord-Hinterland*, 360. It seems more likely that trading parties were being assembled and announced. For more information on the Bali slave trade see Chilver (1967a), Chem-Langhëë and Fomin (1995) in *Paideuma* 41, 191–206 on changes in routes and, in the same volume, Warnier's regional overview. The special case of the Moghamo as a later source of supply for the plantations is dealt with in R.J. O'Neil's Columbia University thesis (1987) and in his contribution to *African Crossroads*, 81–100.

25. Esser seems unaware of a nearer source of salt in commercial quantities – the Mamfe salt springs. Zintgraff had heard of a source of salt to the west of Ntok Difang's village, present-day Tali (*Nord-Kamerun*, 122). The Mamfe source was known to Ramsay of the GNK (*DKB*, 1901, 234–238). For a full account of its preparation and local trade see Mansfeld (1908), *Urwald-Dokumente*, 129–130. These sources were owned by villages that specialized in salt-production and were highly contested. There were also salt-springs just over the Nigerian border (Talbot, 1912, 317). Other than salt from coastal sources, 'black' salt from the Akwana saltings appears to have travelled along the Takum-Fonfuka (Bum) route to northern and eastern Grassfields markets. The Efik at Tom Shott both produced and traded salt from their own and European sources. A few minor sources are reported from the Western Grassfields.

26. For a further account of the *nggwasi*, see P. Valentin (1974) in *Ethnologische Zeitschrift Zürich*, 185–194 and for their elaboration with beads and mirrors see Missionary Göhring's photograph of 1902–5 (in the Basel Mission's photo-archive, E–30.27.017) of royal wives dressed for dancing.

27. 'Bafrum', nearby, is evidently Bafuen or Bafreng, now known by its 'own name' of Nkwen, a Nggemba-speaking chiefdom. According to its oral traditions, it was temporarily scattered by the forays of an earlier Bali Chamba raiding band and its nucleus took refuge in Kom. Nyongpasi or Fonyonga I, who encamped in 'Bafuen' territory before defeating a rival band, the Bali Konntan, and moving to the site of the 'Bafuen', is credited by the Bali-Nyonga with making the return of the 'Bafuen' possible. Be that as it may, their alliance with the Bali-Nyonga preserved a sort of balance of power between Bali-Nyonga and the larger composite chiefdoms of Bafut and Mankon (Bande, Bandeng). Their later iron industry involved the recycling of the slag left by an older smelting industry. See Warnier (1985) 269–270.

28. German military accoutrements – a cuirass and helmet – were later donated to the Fon of Mankon when he had proved serviceable to the Station, and remained in use on grand occasions in the colonial period. For Bamum see C. Geary (1996) in *African Crossroads*, 165–192, and

earlier references cited therein. Max Moisel (Supplement to *DKZ* XXV (15), 1908a) describes the range of uniforms displayed by the Fon of Babungo, another German ally who had welcomed Zintgraff. For a Duala display see the plate opposite p. 186 in G. Ziemann (1908).

29. The assumption that the 'great religion' of the Mubako-speaking group of Bali-Nyonga was influenced by Islam has no ascertainable basis; material influences and military borrowings from Hausa-Fulani sources are another matter. For a study of the practices of the Chamba cluster see Fardon (1990) and for the Pere (Peli, Konntan) component of the Grassfields Bali, see Pradelles de Latour (1995) in *JASO* 26 (1), 81–86. For a valuable early account of some of the beliefs and practices of the settled Grassfields Bali, in particular their 'Banten' (Bamum and Bamileke) component, see a letter dated 22 November 1910 from the Basel Missionary A.Vielhauer to the ethnologist Ankermann. This can be found in Baumann and Vajda's presentation of Ankermann's fieldnotes of 1907–9, *Baessler-Archiv*, N.F. 7 (2), 1959, 271–276. For a Chamba chronology see R. Fardon in *Paideuma* 29, 1983, 67–92.

30. To judge from a photograph of Zintgraff and his helpers of 1891–2 in Dr. Eballa's collection (Yaounde), Zintgraff had acquired the ample gown of a Bali notable and a tufted cap. The photographic evidence is confirmed by an entry in Hutter's diary of 14 November 1891, which records that Bali gowns, sewn on the spot for both of them, were ready to wear at the *voma* festival (*Nord-Hinterland*, 200). He records, too, that on December 6, 1891, at the start of the 'leda' (*lela*) festival he had marched his Bali troop to the market-place and had them fire a six-fold salute. There is a good deal more to this effect in Hutter's reminiscences, indeed more than enough to suggest that Esser's party was confidently expected by Galega and his entourage to put on an impressive show. The abiding love of the Bali-Nyonga for military smartness and competence in drill is well-described by the cartographer Max Moisel who visited Bali-Nyonga in November, 1907 (*DKZ*, supplement to Vol XXV (15) of 11 April, 1908, 270 ff.).

31. Esser is quick to affirm the 'uselessness', for all the party, of Galega's diplomatic gift of four women. Galega had earlier supplied the Baliburg Station with young women, led by Uandi, daughter of the Babessong (Ashong) chief, who seem to have enjoyed some frolicsome times, to judge from Hutter's description of their dancing evenings, with Zintgraff playing German melodies on the harmonium (*Nord-Hinterland*, 178–179). However, Zintgraff had been given the young princess Fé, for whom he had a genuine affection and who was locally regarded as his consort. When Chilver interviewed her in 1963 in her extreme old age and still full of charm she brought out Zintgraff's carefully treasured gifts to her – among them a pair of high boots and a richly-beaded *nggwasi* with a mirror centre-piece, used in dancing – and recalled his concern for her. After his death the WAPV's labour recruiters kept in

touch with her – Kionka (d. 1901) was a special favourite – and brought her little gifts. She was married off to an important retainer, and was looked up to as a person of consequence. In later life she joined the Catholic Church hoping 'that she would meet Zintgraff in heaven'. One of her sons became a WAPV overseer.

32. The trickster-scouts, *bagwe*, are described by Zintgraff (*Nord-Kamerun*, 369) as well as by Hutter in *Nord-Hinterland*, 359, and elsewhere. They are not peculiar to the Grassfields Bali.

33. Esser's description of Galega's bodyguard as 'slaves' does no justice to the status of palace retainers, *tshinted*. For an attempt to recover some elements of early military organization, see Kaberry and Chilver (1961) in *Africa* 31 (4), 360–363.

34. Esser's account of the festal assembly resembles that of Zintgraff when, after his first arrival, in January 1889, and initial negotiations with Galega he was 'shown to the people' (*Nord-Kamerun*, 191–195). The movements described – the three-fold salute or *lö'ti*, war-play and general gun-fire and the instrumental accompaniment are common to the public aspects of *lela*. Early descriptions are to be found in *Nord-Hinterland* at 348–351 and seasonal ceremonies are listed at 429–484.

35. It is unclear what groups were then believed to be 'subject to Bali', i.e. *niwombe* or tributaries. Hutter roughly estimated their total numbers at 6,000. He mentions Mengyen (his Bamingnie) and Bossa and leaves unnamed 'several other' small places (*Nord-Hinterland*, 21, 335). For the original vaguely-worded treaty of 1891, see Chilver (1967a), 487, for Hutter's account *DKB* 2, 1891, 571–578, and for further discussion of the issue see O'Neil (1996).

36. Galega seems to have forgotten his earlier conversation with Zintgraff on the question, see *Nord Kamerun,* 221. An official request made to the Basel missionaries is referred to in ABM Series E2 (1896) 27–28. The mission was then too overstretched to comply, but promised to concern itself with the Bali who reached the plantations. For the first exploratory visit of Basel missionaries in 1902 and their subsequent settlement in 1903 see F. Ernst (1903) in *Der Evangelische Heidenbote,* A. Vielhauer (1936), 45–59, F. Lutz (1907), W. Keller (1969), Chapter 3, and Schlatter (1916), Vol. 3, 279 ff. For a Bali view see M.G. Fohtung, an early Basel Mission pupil, in *Paideuma* 38, 1992, 219–248.

37. For the later history of Tayo, a village or group of hamlets of the Nkokenok II clan, see M. Ruel (1969), *Leopards and Leaders*, 290–292. It is also mentioned in Mansfeld's list of Banyang villages (1908, 265). Maps show it to be on hilly ground.

38. According to Kemner (1922) the resident Europeans, in addition to von Besser, in charge of Victoria Station at that date, and Dr. Preuss,

included the managers of the Woermann and Ambas Bay Trading Company factories, a Government doctor, a police officer and a postal agent.

39. That this may not be an invention on Esser's part, but rather a confused second or third hand report, is suggested by a tale recounted by Hutter in *Nord-Hinterland,* 297.
40. The final paragraph is lifted, almost *verbatim*, from an article by Wohltmann in *DKZ* X, Supplement to No. 15, 1897, 61. Wohltmann explicitly opposed the notion of large-scale German settlement in Cameroon. Esser was not, however, wholly uninterested in the 'emigration question' in different circumstances, as we find him on the board of the Hanseatic Colonization Company, one of Julius Scharlach's enterprises, concerned with German agricultural settlement in Brazil (*Der Tropenpflanzer,* 1901, 385–387).

8

Departure from Cameroon

Once we had arranged in Victoria with Dr Seitz, the acting Governor, to have our concession formally entered in the Land Register, Hoesch and I felt free to repair to Kamerun town [Duala] in his company.

As one approaches it, one first sees the European settlement strung along the river, starting with the Government buildings on the Joss plateau and the civil servants' houses around them, set in a pretty park ... where the graves of the fallen heroes of Germany's most promising colony are to be found. The monument to Dr Nachtigal fronts the Governor's house. The foot of the Joss Plateau is here joined to the river by a quay (built by the Schmidt Company of Hamburg Altona) which has improved cargo handling enormously ... and dispenses with the need for boats. By the banks ... there are still some hulks to be seen at which, in the past, all of the trade was conducted and in which the Europeans had lived. Now the elegant and breezy though massive Government buildings, built during Governor von Soden's time ... and the riverside factories bear witness to German industry and energy!

The town of Kamerun is made up of several villages, which still bear the names of their chiefs. These are the so-called former 'kings', no more than wealthy natives without dignity, caricatures of a real prince like Garega. Their villages ... come into view as one travels further up-river. The factories still own hulks: the river itself is the scene of a lively barter trade. At particular times it is quite covered with boats – the natives' dugouts, Duala war canoes and European row-boats bob about together. Official announcements are even relayed from the river on the drums.

Hoesch fell ill with fever on arrival and was taken at once to the fine hospital run by the Colonial nursing sisters.[1] I was put up in Government House as the hospitable Dr Seitz's guest . . .

Missing no opportunity to do so, I continued studying the character of the Negroes. I found the common view that they reverenced nothing to be generally true. However, there are exceptions. They look up to their parents and normally their love for their mothers amounts to worship. I have experienced this in my dealings with native chiefs: more is to be expected from them if one gained the sympathy of their mothers by means of choice gifts. Gifts to their favourite wives are seldom effective and might even arouse their jealousy. While they may dispose of their wives without much compunction, they will never agree to sell their mothers.

Now for another common European superstition: that Negroes have a peculiar smell and never sweat. Even important scholars have been heard to assert this nonsense. This is total rubbish and if any unpleasant smell assails the nostrils this is due to Negroes wearing unwashed sweaty European clothes, as happens when one meets them in Europe exhibited at fairs and market booths . . . My experiences is that a love for personal cleanliness is general among the Cameroonians . . . as opposed to the Europeanised ones . . . There are, however, some tribes among whom washing is sometimes seen as a breach of custom and who seem to be encrusted with dirt[2] . . . Most Negroes are even vain – another commonplace. They often show excellent taste in dress and their womenfolk can be really elegant . . . Their beauty is of short duration as a consequence of the climate, the heavy workload that most of them bear, and . . . the brutality of their husbands.

[Esser and Hoesch, leaving Zintgraff behind, then boarded the *Nachtigal* for a brief visit to Consul Spengler at Sao Thomé and then proceeded, with great benefit to their health from the sea voyage, via Cabinda, to S. Paolo de Loanda in the Portuguese province of Angola.]

Without doubt, a longer stay offshore is very healthy for everybody who had to live in the tropical climate. It would be desirable to build special ships for this purpose, which offered comforts that we had to do without on the *Nachtigal*. Restored by the sea air we were ready for new undertakings. At S. Paolo de Loanda we were hospitably welcomed by Herr Aengeneyndt in his Villa Germania, and joined by Walter Fournier-Baudach who had come to Africa to indulge his passion for big game hunting and served the later expedition well.

Notes

1. This was the organisation in which Esser's mother was involved. See Part I.

2. Esser may be referring here to the slaves of a group of nomadic herders he met with in S. Angola, the so-called Ba Cubabe (p. 197 of original text), unless he is referring to some mortuary rites from elsewhere, which may be deliberately accompanied by lack of washing and grooming for a period. On the question of African personal cleanliness Esser echoes Zöller (1885, II, 85).

9

Angola, and the Cunene Expedition

[The two chapters about the Angola expedition are not of immediate interest to Cameroon Studies. We have included an abridged translation of them since the expedition was Esser's original mission. It also became the focus of the 'Esser Affair' of 1899, for which see Appendix I, and contributed to Esser's public (if very temporary) disgrace. They may also contribute to some understanding of his character which was not without the daring and jesting he observed in the trickster scouts, the Bagwe, of Bali-Nyonga.]

The Angola Expedition

[After Esser left Kamerun he first stayed briefly in Loanda, where he was well received by the Viceroy.]

The task that I had set myself was to study the situation in Angola in order to be able to compare the latter with German South-West Africa. This task may appear foolhardy given the differences between the two countries... Despite of the great extent of the two possessions and without applying generalisations to the whole of them, one should be able to compare southern Angola with northern German South-West Africa, which are only separated from each other by the Cunene river. One could try to conclude in what respects 300 years of development of the Benguella and Mossamedes Districts could reflect the future and development of the northern part of German South-West Africa.

The Angola Province has been and still is of greatest importance to Portugal. During the last financial crisis the islands of Sao Thomé and Principe, as well as the Angola Province, have saved their mother country Portugal from complete national bankruptcy, not without themselves suffering badly. These African possessions, which required a lot of capital input in the course of the centuries, have been able to maintain the state and do so currently today.

[A short geographical, geological and climatological survey of Angola follows, with passing references to the nitrate and guano deposits exploited by the Mossamedes Company and its use of seasonal rivers for the irrigation of plantations. Mentioned – and the theme recurs – is the use of convict deportees, let out of prison for harbour and road labour, and later freed on bail or else given a parcel of land to work, with their families, who were given free outward passages. Esser claims to have conversed with deportees, many engaged in cattle rearing, and concludes that their contribution had been very useful. The independent Boer settlement at Humpata is mentioned. He turns next to the Holy Ghost Fathers, mostly German-speaking Alsatians, and to their economic activities.]

The Holy Ghost Fathers, whose mother church is in Paris, have established missions in the whole of Angola. Their influence has been beneficial to the natives for centuries, but mainly where the missionaries work themselves and have employed and educated the Negroes. They employ 900 workers growing sugar cane in Huilla, and, apart from growing sugar and coffee, they had contributed to stock-raising and . . . had introduced vegetables and fruit from all over the world . . . The missionaries should be credited, first and foremost, with having created the existing roads. The Fathers had already started to construct roads 200 years ago . . . Curiously, nearly all the Fathers are German Alsatians with real German names. I greatly preferred these German men at Cubango to the over-zealous English and American missionaries.

Only one of the five railways shown on the official maps exists. This line runs 322 km from Loanda to Ambaca and is used to transport the riches of the hinterland to the coast at Bengo. A small tramway of 20 km serves the coffee groves and the trade centre of Catumbella. Another line is projected from Mossamedes to the Chella range where 8000 settlers live.

[A rather confused account of the native trade follows – cattle were bartered; various natural products – feathers, hides and skins, wild rubber and other forest products, such as wax, ebony, gum-arabic, camphor-wood, were traded, as were palm products.]

Two ships a month . . . carried produce to Lisbon. The riches of the country also reached South Africa. The Swedish trader Erikson is said to be the biggest trader in the country and to arm whole tribes to kill elephants and ostriches, to come with Boer companions from the Cape, with wagon-loads of trade goods, to stay for some time, and to return with his booty of ivory and feathers to the Cape . . .

In the hinterland of Loanda, Benguella and Mossamedes, coffee trees grow wild. The Portuguese planters led by the *Banco Ultramarino* have taken advantage of this coffee land by burning down miles of virgin forest and have made over extensive regions to coffee. Once sown, nobody gives any further care to these groves. Factories, that is all, have been constructed to which the natives bring the beans that they have collected. They are allowed to pluck the beans and are compensated for each pound of beans they deliver. The coffee is unselected and only of second quality: this cheap investment, however, provides a good return. It was said that not up to a fifth of the coffee grown entered the trade . . .

Native-grown crops are maize, sorghum, tobacco and long-staple cotton, the last of good quality . . . The Europeans mainly grow sugar cane, profitably converted into spirits for local use as import duties were high . . . Their labourers were really slaves, like those of Sao Thomé . . . and very ill-paid. The [Catholic] Mission paid them more and treated them better . . . I have both witnessed the economy of the slave system and been told how it works . . . Man-hunts were conducted by the Ballundo [*sic* Bailundu] and Banano tribes who brought their captives to an estuary near Novo Redondo, where they were sold at 200–400 Marks a piece to Portuguese traffickers. They were sent by sea to Cabinda, Loanda, Benguella and Mossamedes and despatched inland. Their 'contract' was officially stamped, prohibiting export to non-Portuguese territories . . . No landowner pretended they were not slaves.

[Esser then turns to colonial customs posts and to widespread arms smuggling.]

An incident that happened during our expedition from Mossamedes via Cap Negro while moving up the Coroca is significant. Near to the river we met a troop of wandering Damara who were willing to exchange 12 oxen for gunpowder and ammunition. Asked what it was needed for, they proudly declared that they were fighting a battle with a powerful German chief, Major Leutwein, and that they had shot his white horse from under him, killed two other officers and wounded another two. Now they had run out of ammunition and had had to draw back. You can imagine that I felt like imprisoning them but, apart from being on Portuguese territory, they were too many of them compared to our small expedition . . . We were convinced that they were eventually provisioned by Portuguese traders in Humbe . . .

Of the Stations shown on the map of Mossamedes Province some had been destroyed, and some had never existed . . . A band of Boers kept a border watch at the confluence of the Elephant River with the Cunene to prevent regular cattle-raids by Hottentots armed with breech-loaders... The local under-equipped forces had failed to do this...

The provincial government had been alarmed by our expedition to the Cunene and the Tiger Bay. They wanted to keep recent gold finds to themselves and avoid an influx of foreigners who might edge out the Portuguese... Eight years earlier Portugal had offered Germany a strip of the Cunene area because at that time the raids by the Hottentots were particularly numerous and the Portuguese were unable to fight them. The offer was ignored... but perhaps there were fears that it might be revived.

> [Esser then describes, as a comical event, the reaction of a Portuguese gunboat in Tiger Bay when Esser raised a German flag above his tent, as was his wont. The local press described the gunboat's reaction in glowing, patriotic terms.]

When we had arrived in Tiger Bay, and our black-white-red flag waved in the wind from the top of our tent... the gunboat Duoro fired three times... thousands of flamingoes rose to the sky... Immediately after this, there landed a party which came running towards us... and finally fired into the sandhills... Next day we were invited to breakfast on board the *Duoro*... Nobody talked about the flag. Just one junior officer warned us that the shore was mined... I later had an interview with the Governor who insisted that I had hoisted the German flag at several points on the Cunene on Portuguese territory – an untrue contention. He refused to believe my explanation, and announced his plan to go there himself to see what we had done. I explained to him that my expedition was a fact-finding one, which had the support of the Mossamedes Company. The next day we learnt that our interpreter had been arrested. The Governor explained... that our assistant was not allowed to leave Portuguese territory without a permit and that he had been with us – which was true – on German territory when we went back along the Cunene to the Augusta-Victoria-Bay that I had discovered... Two days later, the Governor went off with a large following of camels, oxen, boys and women, up the Cunene... I doubt that he went far with his big crowd. He will have found neither German flags nor treaties with chiefs. We had neither left nor made any...

> [The next Chapter supposedly deals with the expedition itself. Passing references to German South-West Africa are omitted, and this part restricts itself to what can be learned about Esser's route.]

As mentioned before, after our arrival at Loanda we were kindly welcomed by the Viceroy, Mr. Guilhelmo de Capello. We stayed briefly in Benguella where we received information from two German residents, Peters and Schäfer. Next we were welcomed at Mossamedes by the Company and made preparations for the expedition, the aim of which was to discover whether the Cunene river could provide a supply-route to the north of the German colony. The entire German South-West Africa has a savanna and desert climate . . .

Departing from Europe, Mr. Hoesch and I had at first not intended to head expeditions ourselves. We had given the lead to the experienced Dr. Zintgraff who, as we know, remained at Bali in the hinterland of Cameroon. Hoesch and I, irresistibly attracted by the rich hunting grounds, now had to proceed on our own. Since we had not planned for this contingency we had not prepared ourselves in survey skills. It is my duty to declare this with respect to the sketch-map for the provision of which we had to rely on such knowledge and experience as we had acquired during military service.

The Portuguese have with the excellent harbour at Mossamedes and despite their small investment of money and labour, created one of the richest ports in a short time. At present, this promising place already maintains 3,000 Europeans . . .

The expedition consisted of Hoesch, Fournier-Baudach, who had been in Angola before, Sampaio Nunes, the Portuguese interpreter, myself, 55 carriers, several camels and pack-oxen to ride on. The camels were loaded with firewood and water for the dry wilderness. We could not get used to ox riding and preferred to walk like the carriers . . .

From Mossamedes, we crossed the dry wilderness . . . In two days we were at Port Alexander . . . which is the source of much of the salt fish beloved by African labourers.

[Esser then gives an extended account of the fisheries.]

From here, we followed the dry bed of the Coroca . . . flanked by sugar plantations ingeniously irrigated in the dry season by distillery pumps . . . From the Coroca we went through grasslands to the cave-dwelling Makua Matafe, who fled . . . In a valley running from north to south we met a herd of over 30 elephants, 3 of which we shot and which were consumed by our carriers in 15 hours flat.

[Now follows a description of stomach massage to aid digestion identical to that performed by their Vai carriers on the Bali expedition.]

A few days march further, on the South flank of the Chella range, we encountered Muquich nomadic herders, brave, intelligent folk, and their slaves, the Ba Cubabe. The latter were the stupidest, dirtiest, Black folk I had ever met, wretchedly disfigured by diseases and living in filthy huts ...

On the descent of the eastern slopes of the Chella range, lo and behold, a very different group, the Myi Buba, proper Hereros, and the Mundimba, who resemble the Mum Humbes, the princes of Humbe. These were intelligent, brave, good shots, clean, 'free sons of the highlands'.[1]

[There follows a passage describing their musical instruments, dancing and rain-makers in some detail.]

We heard of gold finds by an English expedition north of Cassinga and north-east of Humbe but had no time to visit them. That the region is gold-rich was confirmed by our meeting, on the River Jabo, a band of Boers ... with their dressy wives ... and their slaves, the latter washing gold in river dams under the supervision of their masters who carried long whips of giraffe leather to pick out and punish the sluggards ...

The first discoveries of gold in the region had been made by a German, Arndt, and two Dutchmen, the brothers van der Keelen, who had detected geological formations like those of the Transvaal ones ... The presence of alluvial gold was later confirmed by the Catholic Mission in correspondence.

A day's journey took us to the banks of the Cunene [which is lengthily and repetitively described with its narrow gorges and cascades in the upper reaches and then widening and embracing islands rich in wildlife].

The English seem to be well informed about the rich wild life. We met at least one hunting party of 5 men ... From the cataract of Quenguari onwards both sides were thickly populated: the Ohinga branch of the Ovambo were settled on the German side. The breadth and depth of the river varied seasonally ... leaving both swampy and rocky patches ...

Before the falls at the Black Mountains were reached ... the river ran through an easily bridgeable gorge, then broadened to include islets ... At the Elephant River inflow we met Boer ivory hunters ... At the St. Mary River junction we could see the sea from a hilltop ... I believe the Cunene is navigable from here to just before the mouth, and I found no trace of ... waterfalls of 40 meters as shown on Portuguese maps ... When we descended along the Cunene, we saw that 200 metres high sand dunes, starting at the St. Mary River, prolonged around the eastern slope of a 20 km long rocky plateau in a south-eastern direction. The river's exit however continued to the north of the rocky plateau. We thought we detected traces of a dried-up stream bed, perhaps a second river exit, which we followed over the border. We came to a small bay . . . It was fairly calm and approachable only from the north but it became very rough when a storm blew up . . . We named the bay after the Empress Augusta Victoria . . . Even if it was difficult to land in the bay, it was worth further study as a potential haven for commercial and strategic reasons, as it was within reach of the navigable stretch of the river . . . The outlet of the Cunene is blocked by sandbanks and is very shallow for two kilometres upstream . . .

The distance from the valley of the Cunene to Tiger Bay, we discovered, is much shorter than that shown on maps and there are no intervening hills . . . Nor are Tiger Bay's dimensions as depicted on all available maps . . . It is not blocked by sand. The Tiger Bay is the largest natural harbour in the world. Two Portuguese naval vessels were manoeuvring in it without difficulty and it could have taken many more . . .

I am convinced that a railway joining Tiger Bay to the navigable part of the Cunene would be rewarding. It could stand as the first leg of a trans-African railway along the Cunene, across the territory of the Mossamedes Company, of the *Kaoko-Land-Gesellschaft* and of the Chartered Company up to Transvaal and Pretoria, ending in the Indian Ocean.

Whether the Augusta-Victoria bay that I have discovered will one day turn out to be useful or not, it will always remain of minor importance compared to Tiger Bay. It was of little importance to Portugal – Angola being rich in ports anyway. The German government is highly recommended to keep an eye on this excellent port and acquire the Tiger Bay area, wholly or in part. Before a large-scale immigration to the left bank of the Cunene can take place, the German flag has to wave a friendly welcome to the ships.[2]

Notes

1. Compare Esser's description of the Bali.
2. Esser is totally silent about the earlier exploration of the Cunene banks by von Hartmann on behalf of the South-West Africa Company and their railway plans. It seems unlikely that Esser could have been directly informed of any secret discussions prior to those of 1898 for a division of the spoils between Britain and Germany should the Portuguese Empire

fall apart for lack of resources. If that had been the case Germany might have acquired territory or a sphere of influence in the centre and south of Angola. The Portuguese Empire did not break up: it had been prevented from laying claim to lands which would have joined Angola to Mozambique by a combination of British and Cape interests. The British found their old alliance useful when the Boer War broke out in October 1899 and guaranteed the integrity of Portuguese territories. See Warhurst (1962) and W.R. Louis (1967) in *Britain and Germany in Africa*, 23–28, 36–39; also J. Willequet, 268–269 for the much later participation of German capital in the Angola railways, the initial reason why Esser's trip was commissioned. The notion of some exchange of territory to benefit German South-West Africa, however, never seemed quite out of the question; see Konsul Singlemann in *Kol.Rundschau* 5, 1912, 272–280; see also Barth (1995) 64ff., 410ff., and Drechsler (1996), in Chapters 2 & 3.

10

A Retrospective View

With the study of Tiger Bay completed, our task was fulfilled. We hired a sailing ship to Mossamedes where we were warmly greeted by the Company's officials and thanked for the geographical information we handed over concerning their property. Our travel on foot was acknowledged, and we were told that after our decision not to ride on the oxen, nobody had believed any longer in the success of our journey, since in Portuguese terms such distances are impossible to overcome on foot. Yet, our travels in Cameroon had certainly been more strenuous than the Cunene expedition. In the course of both expeditions, we had covered around 1500 km on foot. I cannot deny that at our arrival in Mossamedes I had had enough of trekking. On the average, we had covered 42 km a day, from 5 a.m. to 8 p.m., with only 3 hours rest.[1]

Arrangements kindly made by the Company for us to sail by a French steamer to Cape Town unfortunately fell through, so we gave up the idea of reaching the Cape and decided to join the next homebound Portuguese ship, the *Zaire*, via Loanda. We spent a little more time in Mossamedes where we went on hunting excursions. I had my interview with the Governor, already described ... and used the time available in studying the possibility of using deportees ... as farming and cattle-rearing settlers, after a few years' work as roadmakers and port labour ... on the left bank of the Cunene.

[The next four pages deal with a question debated solely in relation to German South-West Africa. Esser generally favoured the deportation of convicts as both humane and economical. He quotes chapter and verse from a work by the jurist Dr. F.F. Bruck, from his published controversy with Graf Pfeil, and an opinion by one Dr. Fritz Freund of Koblenz. Esser cannot resist slipping in the prospects of a livelihood from wild rubber collection on the Cunene banks.]

During our three-week stay in Mossamedes, and in the course of a hunting expedition . . . Hoesch drank some bad water and picked up dysentery, which later involved a long stay in hospital in Germany . . . This made a rapid return to Germany more urgent.

We boarded the *Zaire* in harbour and spent two days in Benguella while the ship took on coal. Here I visited my acquaintances Peters and Schäfer. Peters told me that he had bought a large diamond from a Boer and had already sent it to Amsterdam to be valued. He also knew where it was found, so we concocted a plan for a future expedition from Benguella into the interior up to the Zambesi, to reconnoitre the region for occurrences of precious metals and precious stones.

The *Zaire* reached Sao Thomé via Ambriz, Ambrizette and Cabinda. We were met in Sao Thomé by Spengler who had, meanwhile, been in Victoria, informed himself on the planned plantations and was pleased with what he had seen. For me, this was a reassuring report from an expert. Here, too, the *Nachtigal* called, enabling us to send news of our return to Zintgraff and Preuss.

Fournier-Baudach had been left at the Cunene river from which he went to Ondonga and Kwanyama in German territory, but his plan to trek to Windhoek, as he wrote, was prevented by rinderpest regulations, so he re-crossed the now rising waters of the Cunene.[2]

From Sao Thomé Spengler travelled back with us on a visit home and to be at our side during the setting up of the plantation company.

From Principe, the next stop . . . we sailed to Praia in Santiago [Cape Verde Islands]. During this trip, two passengers, who had died of fever, were buried at sea . . . At Praia, I went ashore to shoot game birds, with an English engineer and a Polish Professor, and brought our booty . . . on board to be cooked, to the joy of the passengers. Then we stopped at Madeira and finally landed in Lisbon. I found an old friend, Kleist, in my hotel. The next day, I had a long interview with the German ambassador, von Derenthal, chiefly about the Portuguese colonies, and was kindly received by his family. After four days we boarded the *Süd-Express* for Germany.

An interesting part of my life lay behind me, during which I had been constantly learning. I am convinced of Cameroon's fertility. I also think that the unsurveyed parts of German South-West Africa are no worse than those of their Portuguese neighbour, where thousands of men get by without pumping works. When assailed by thirst our caravan would have found water by means of pits and boreholes. The riverbed of the Cunene provides a permanent reservoir, as the waterholes of the natives bore witness. There are mining prospects in the northern areas of the protectorate should the finds on the Portuguese side turn out to be worth developing. However, stock-raising would offer a real future despite tse-tse fly and rinderpest. I hope that my description of the Portuguese side ... will provide a basis for profitable investment by German enterprise. Once the first half-dozen colonists have earned a living for themselves and others from the soil, and once a couple of thousand cows and sheep have watered on the German bank of the Cunene, further development will follow naturally. The rewarding soil of the Cunene only calls for German industry and efficiency.

That the conditions for a plantation development in Cameroon have been demonstrated is due to Professor Wohltmann of Bonn, to whom we owe the solid beginnings of agriculture and plantations in the German colonies ... That the colonial partisans have not striven in vain is well-known to other powers, as both the jealous English and more open-hearted French have shown in their different ways.

The latter have shown it in the untroublesome agreements over Togo and in their colonial literature.

Professor Leroy-Beaulieu, the foremost colonial historian of the times, in his *De la Colonisation chez les Peuples Modernes,* 1890, 315, wrote as follows:

> 'The Germans have given notice that they are able to colonise; they have plenty of capital, which they lacked fifteen years ago; they show an outstanding spirit of enterprise; one can expect their success, which should not be begrudged them.'[3]

Leroy-Beaulieu was convinced that the task, nowadays, of an aspiring, healthy and capital-endowed nation was permanent colonisation. Other nations, such as Italy, Denmark, Norway and Sweden had lost their opportunity but France, after its unhappy defeat in 1870–1871, saw that the only opportunity of remaining a first-class power open to it was the acquisition of colonies. It was by good luck that Germany, even if out of instinct rather than because of the insights and foresight shown by France, was not left behind, but became a colonial power.

The most necessary thing is for German capitalists to pluck up courage and turn their attention increasingly to the colonies. The German people spend five hundred million Marks a year on colonial products; how much of this could remain in German hands! All that is needed is some entrepreneurial spirit and a little patience, since the development of the plantations being established naturally takes time. It is my firm opinion that after five or six years the capital invested will show excellent returns. It has often been stated that our new efforts will drive the Fatherland into world politics and that Germany should have nothing to do with them, just as it was maintained by most, fifteen years ago, that Germany had no interest in colonies. Whether or not we engage in world politics [Weltpolitik] is not a matter for decision by individuals but determined by our national economy. And if this occupies a leading position in the world economy, one of the first after England, if not the first, it would become impossible to avoid taking part in world politics and imperative to engage in them further.

Germany has already once lost its position as a world power that it had held in the Middle Ages as a result of unfortunate internal quarrels and its own failings. We Germans could not face the judgement of later generations if we did not now, at the right time and under the strong and purposeful leadership of the Hohenzollern dynasty, abandon our ancient narrow-mindedness and short-sightedness and restore the grandeur and might of the German Empire, and allow the German spirit to unfold anew.4

Notes

1. Esser's calculation of average daily distances might well have been sufficient to raise doubts in the minds of experienced travellers; Fournier-Baudach had attempted to explain this calculation by stating that the Boers had built a road from the Chella range to Humbe which had speeded up travel, in a letter to the *Tägliche Rundschau* dated January 30, 1899.

2. Wagner tried to make use of an 1896 contribution by Fournier-Baudach to a hunting journal, *Wild und Hund*, to demonstrate that Esser's expedition had not travelled any great distance from the coast and to sneer at the risks it encountered. Fournier-Baudach felt obliged, in the letter quoted above, to reproach Wagner for his 'frivolous and unqualified' assertions; Fournier-Baudach himself had declared that he was partly responsible for the map of the itinerary of the 'expedition' and had checked it again on his return journey.

3. The French Professor of Political Economy had also remarked that the Germans had found the climate of Africa more taxing than had the English and French.

4. Esser's peroration includes many of the commonplaces of the 'Colonial Movement'. See, for example, a contribution signed F. Sp., published in Supplement (May 1) to the *DKZ* X, 1897, for resemblances in the rhetoric, in addition to Wohltmann's contribution of the same year, in

Supplement no 15 (June 12) to the *DKZ* X, 1897, which Esser had already laid under contribution. His unnamed adversaries might range from those conservative agrarians who regarded the new colonies as a drain on resources which should be devoted to the maintenance of German security in Europe, in particular against Russia, to members of the Opposition in the Reichstag, from the reformist wing of the Centre Party to the Social Democrats. The 'Colonial Movement' itself was composed of different groups which ranged from the *bien-pensant* National Liberals – see chapter 13 in Part III – to the populist chauvinists of the New Right with a racist mystique. See Austen (1986) and the references he quotes, in particular to Eley (1980) and Chickering (1984).

If the term *Weltpolitik* has been explained by Stengers (1967, 343) as 'a vague but eager wish to equal Britain throughout the world' in its more popular expressions, Esser's use of it refers, rather, to the unfolding situation in German foreign policy in the mid and late 1890s. This is conveniently summarised by Woodruff D. Smith (1978, 169–179, and 1986, 52–82) and reviewed by V. R. Berghahn (1994, 220 ff.) in its domestic aspects. The entanglements of foreign and defence with domestic policy have remained a central topic of German historiography since the 1960s – see, for example Pogge von Strandmann (1969), a provocative study by H-U. Wehler (1973, transl. 1985), and Förster, Mommsen and Robinson (1988) for its earlier stages. Esser's particular conformities, preferences and prejudices, including his Anglophobia, are not unexpected given what we know of his background and the connections he inherited or sought.

Figure 10.1: The young Max Esser with Victor Hoesch and others in Cameroon: an early undated print from his daughter's photograph album, which was made available by courtesy of Herr Josef Pfaffenlehner, Schloss Sandersdorf.

Figure 10.2: WAPV beginnings in 1897 at Kakaohafen: from Kemner's pamphlet of 1922.

PART III

Colonial Needs and their Consequences: the Viewpoints of some Contemporary Observers

11

The 'Bali Road' and Baliburg in the Autumn of 1892: a Report on a Visit: Max von Stetten[1]

This report, published in the *Deutsches Kolonialblatt* (4, 1893, 33–35), was written four years before Esser set off, side by side with a report (pp. 36–38) on the Bali district by Lt. Franz Hutter, dated September 10, 1892, in which the Bali are compared with the ancient Germanic peoples in their habits, and depicted with affection and sympathy. How far the report we have translated represents von Stetten's own views faithfully is far from certain: it has been selectively edited in the Foreign Office and presented in the third person.

Zintgraff left Baliburg in June 1892 to reorganise the intervening Stations, leaving Hutter in charge. Hutter's own diary of events is to be found in *Nord-Hinterland,* 7–27, and makes no more mention of von Stetten's visit than the latter apparently does of Hutter's presence. On 1 January, 1893, Hutter received orders to vacate the Baliburg Station and close down the rest. Two months later the Zintgraff expedition was formally dissolved, and recriminations followed.

We have selected this report to illustrate the conflicting views held by the German authorities at different times (and sometimes simultaneously) of the Bali, according to the uses it was proposed to make of their manpower, influence and attachment to the Germans, in accordance with policy fluctuations and budgetary constraints.

1. The forest country

Von Stetten arrived at the beginning of September 1892, in Mundame on the Upper Mungo. The Station, under von Steinäcker, was very well developed. Apart from him the caravan-leaders Wisotzki and Carstensen were present with 15 of their people and a few artisans. Work on its surrounding provision-garden is effected by Balis. There is a wooden house and corrugated iron barracks. A storehouse is under construction. The site is in a clearing planted with palms, cocoyam, manioc, maize and rice and, in addition, potatoes and other European vegetables. Meat is only to be got at very high prices. The principal value of the Station is as a base for caravans. Von Stetten took 14 days from here to reach Bali and 11 days to return.

The population along this stretch is generally agreed to be fairly dense. Starting from Mundame we meet the following tribes:

(i) The Bakundu from Mundame to Kombone [a Bafo village of the Badjuki clan]. Villages on the caravan route were Mokonge [the Balong village of Mukonye], Mampanda [Mamban or Mambanda, a Bafo village], Kiliwindi [or Ikiliwindi, a Bafo village of the Bapewan clan], Bafumati, Mbullu, Bakundu ba Konge, [Bakundu ba Konye], but all but the first three were deserted.

(ii) The Batom [a Bafo clan name] from Kombone to Mabesse. Villages on the caravan route were Nunga, Siba, Pundediko, Batom, Kokobuma, Dipenya, all deserted up to Kokobuma, Batom and Dipenya.

(iii) The Mabum or Babum [possibly Bebum, a Mbo-Miyengge group] from Mabesse to Nguti (Sokwe). The intervening villages of Fobia [a Mbo-Miyengge chiefly title] and Bakuni [a Kenyang-speaking enclave] were deserted.[2]

The Nguti speak a language like one of the Banyang ones [sic!], at least the characteristic sounds of this language are recognisable. As to the character of the inhabitants of this zone they seem little inclined to war and are very industrious. They seem to be well-endowed as traders. The Kumba [then Bafo of the Badjuki clan], Kiliwindi and Mokonge folk engage in trade with the factories and can be met with in groups of 8 to 10 going out on or coming back from a trading trip. The inhabitants of particular villages have their chosen outlets. The Kumba and Kiliwindi folk trade with the Jantzen and Thormählen firm's factories in Mundame and Kiliwindi, there mostly in rubber; palm kernels and camwood are mainly taken to the Rio del Rey and, moreover, 'English' middlemen come in from the west to Barombi. The Mokonge folk are the trade-friends of the Duala who work with English money (trust and commission) and are to be found in the factories along the Mungo up to Mundame.

The trade of these places does not extend beyond Batom. The Batom folk themselves trade with the Banyang. In Batom, the trade becomes divided into that from Mabum which is directed to either the Calabar (Cross) or Mungo rivers, and that of the Banyang which goes direct to the Calabar. The main article of trade is rubber, which comes in inexhaustible quantities from Banyangland and, in addition, palm kernels and ivory. The area between Mundame and the Calabar river is remarkably rich in elephants.

Forest and bush alternate with extensive cultivated lands, the latter unfortunately largely abandoned and overgrown, but bearing witness to the fertility of the soil. Plantain and cocoyam plantings are most frequent but maize, manioc, yams and groundnuts are also grown. In Nguti fine pineapples were also offered for sale. Oil and wine palms are to be found in quantity everywhere and coconut palms are to be seen in Kokobuma.

As to livestock, one meets with sheep, pigs and goats and, in Mokonge, Kiliwindi and especially Kokobuma fine cattle as well. Were an intermediate Station to be built in the forest country the high prices paid in Duala for cattle suggest that cattle-rearing should be undertaken by it.

The prices paid in the forest area vary considerably and reflect the distance of places from the waterways. So while one can buy a half to 1 lb ball of rubber in Nguti for 4 or 5 leaves of tobacco, this would fetch 8 leaves in Batom and one 'head' in Kiliwindi. In Nguti one goat fetches 3 fathoms of cloth and one is charged one leaf of tobacco for 4 eggs or 5 leaves for a fowl. In Kiliwindi a sheep or goat fetches 6 fathoms of cloth and a fowl 3 'heads' of tobacco. In Mundame the cost of victuals for Europeans is so high in barter-goods that it is easier to pay for them in ready money.

The barter-goods that do best up to Bali are good, not light, cloths – poorly printed cloths are not willingly taken – black and white beads, and, above all, small red beads. Tobacco is no good as a means of exchange beyond Tinto. In addition to these, though in lesser quantities, are small snuff-boxes, brass chains, china buttons, nails and, of course, gunpowder and guns. From Batom up to the Banyang, one mainly sees the long English flintlocks. The natives have almost all drawn back from the caravan route because the Balis going back and forth do not respect their property at all and are given to all sorts of violent acts.[3]

2. The Banyang

The Banyang are to be found from the rivers north of Nguti – that is if the Nguti are not to be reckoned as belonging to this tribe – up to Sabi [Sabe's, or Sabes on recent maps]. In between lie the Tamba villages of Difang, Fotabe, Tinto and Miyimbi.

A Station has been established at Tinto that is very picturesquely situated and commands a view of the Cameroon Mountain range. Here were to be found the caravan-leaders Ehmann and Goger with 25 of the expedition's people and Bali labourers. There is a mat-walled house built with stakes. Work here only started a short time ago. The soil is good and meat sufficient and available cheaply. Relations with the surrounding natives are peaceful and they come daily in great numbers to trade at the Station. The aim of the Station is in particular to maintain the trade opened by the expedition with the Banyang and to control the Balis on their way down. The population density seems to be high, since to judge from inquiries a large number of Banyang compounds lie for kilometres along the side of the route and, on the route itself, between Nguti and Tinto, at least 30 were passed. The Banyang have no high chief but are divided into a number of smaller communities. The hitherto most powerful chief was Difang who lorded it over the northern half of the area between Otombe and the river of Calabar. But recently Difang, after he had fallen upon his fellow-tribesmen further south, was decisively defeated by them and fled northwards: all his compounds are either in ruins or abandoned. The chiefs of Sabi and Miyimbi now seem to be the strongest, but there seems to be no cooperation between the Banyang villages, a cooperation called for in the interest of their self-preservation against the Balis going down to the coast as well as desirable for us, since it would provide a real defensive barrier for the forest country. Among the Banyang the Bali comport themselves with more restraint than in the forest.

The Banyang are a more self-confident tribe than those of the forest and outstandingly industrious: they were forthcoming and friendly. The area is a real garden, alternating with their tidy compounds. All the above-named forest produce is grown here in well laid out farm-plots. Oil palm groves reach to the peaks of the foothills. There seems to be an unending source of wild rubber here. Goats, sheep and fowls are present in quantity and wild partridges and pheasants are to be met with. There are large numbers of elephants. Trade consists principally of rubber and ivory, as well as palm kernels to the west. This trade is handled by the Miyimbi Station of the Jantzen and Thormählen firm which makes use of both Nguti and Banyang carriers and uses the Nguti on the southward route.

Five days journey west of Miyimbi [present-day Tali, now divided into several quarters and villages] there is a white English trader, indeed on the banks of the river of Calabar itself, which is navigable there according to the natives. Earlier the Station used to give 6 china buttons for a ball of rubber; nowadays they give 15, or a fathom of cloth for 8 balls. Apart from this prices range as follows: 1 goat = 5 to 6 fathoms of cloth; 1 pig = 4 to 6 fathoms; 1 fowl = 1 yard to 1 fathom or a spoonful of small red beads; 1 large bunch of plantains is exchanged for 4 leaves of tobacco or 1 yard; a large yam, or 3 eggs, or 8 maize cobs or 1 bowl of groundnuts can be got for a leaf of tobacco.

The Banti people are found to the north of the Banyang. The next people, the Bamessong [Ashong], are completely dependent on the Bali, and wholly attached to them and were armed by Dr Zintgraff for this reason. All comings and goings to Bali must cross the steep Bamessong hill, which is why they hold the key to Bali.

3. The Bali

The village lies along a lengthy ridge surrounded by numerous farms and slave settlements.[4] The number of men of fighting age is estimated at 2500 and 3000 with sworn allies. The houses are built of clay with pointed grass-thatched roofs and, together with the forecourts and streets, are in general kept clean. The Station is unfortified. It lies on a ridge parallel to that on which the village stands.

On his arrival von Stetten received from Garega a greeting gift of 3 goats and palm wine, incidentally the only gift received during his 15 days stay. The traffic between the Station and Garega and the Bali was undertaken by two notables, Fomaku, a younger son of Garega's, and a trusty, by name Fonte. The latter spied round the Station all day. Garega received as gifts 1 Hausa robe, a piece of brocade, 2 Hausa daggers, rings, bracelets, chains, necklaces, earrings, beads and nails. These gifts do not seem to have pleased him, for next day he asked for the usual cloth. According to the accounts of his companions von Stetten had expected to find in Garega a dignified elder, similar to the black Moslem princes, but found instead an easy-going, sometimes childish old man, endlessly quaffing huge quantities of palm wine.

He was very friendly and made clear with commendable open-heartedness that he liked the Whites so well because their presence had brought wealth and prosperity to his village.

The Bali are divided into three classes – freemen, dependents and slaves. Only the first, those who have maintained their origins intact, present a handsome, tall, intelligent appearance and in social intercourse are altogether more likeable than the Bantu of lowlier status. The relations between slaves and their masters seemed to be very lax. Slaves can travel without permission from their masters to the coast and deliver a part of their earnings to them when they return: that is all that is required of them.

The effort of the whole people is directed towards acquiring, in the shortest possible time and with least effort, the treasures of the Whites. They beg for them in the most shameless way. The well-known and experienced Caravan-leader Cornelius confirmed that he had never before come across such a covetous people.

Since a great quantity of cloth reaches Bali it can be observed that perhaps two-thirds of the 2000 men present are clothed: many wear trains two or three meters long which were naturally spoilt and torn in the course of a day's wear.

The Station itself is supplied with 50 Balis – slaves – as soldiers and workers, and five women. These soldiers belong to the Station and are victualled, clothed and paid for by it. These poor folk have the existence of the Station to thank for their better situation.

Their farms, in general well-tended, in which they cultivate cocoyams, yams, maize, groundnuts, sweet potatoes etc. provide their sole subsistence, otherwise they have nothing noteworthy to offer since very few elephants are killed in the region. This is the reason why their movement towards the coast, which Dr Zintgraff wishes to promote artificially, is irresistible. They have already gone in parties on the march towards the coast to buy rubber etc. against beads and so to function as middlemen in reverse.

Farm-work is performed solely by women and slaves. The price of foodstuffs, given the distance of the Station from the coast, is not low. The currency is the brass, a coil of brass wire, which is about a yard long when extended. It is worth a yard of cloth, a handful of small beads, or twenty large beads. A goat costs 5 to 6 fathoms of cloth, a pig up to 9 fathoms, a fowl costs 1 yard, a bunch of plantains half a brass, and 10 eggs cost a yard of cloth.

Von Stetten did not think that there were great prospects for trade from Bali to regions to the north and north-east, where much ivory was reportedly to be found: the Benue waterway was too near and traders from there could not buy goods more advantageously here. Actually Hausa traders come up to Bafut on horseback a few times a year.

He reports further, on whether the Bali, who do not produce any exportable goods, can be converted into carriers, labourers or soldiers.

He remarks that his acquaintances are unanimous in the opinion that they are useless as carriers, thievish and unreliable; he himself was able to convince himself of this by experience, on his return journey. This view would appear to be shared by Dr Zintgraff himself since he recruited his band of 30 carriers, who ensure the movement of goods within the group of Stations themselves, from abroad. Actually, without such a permanent column of carriers it would be impossible to maintain the whole enterprise.

The postal traffic between the Stations is undertaken by the Vai folk of the Expedition as the Bali have not proved reliable.

There have been insufficient opportunities for observation on which to base accurate advice on the Bali as labourers. The work that they have so far been engaged upon on the farms of the Station is mostly light and not intensive enough to satisfy an employer requiring disciplined daily task-work. Given the rather unbiddable character of the Bali it might not always be possible to find the appropriate supervisory staff. Moreover the Bali would not willingly commit themselves for more than a few months. Von Stetten suggests that if one wanted to test their value as labourers so as to arrive at a final view, 60 to 80 men should be engaged for one year as plantation workers, say for Victoria. Yet it might be difficult to obtain this number of people for a whole year.

As to their employment as soldiers, he does not doubt that they would fight bravely, and that, if it became necessary, the temporary support of a Bali auxiliary force would be helpful given some previous training. But he could not recommend that they should form a permanent section of a professional Protectorate Force as he did not regard them as amenable to strict discipline.

In conclusion he describes how, because of a shortage of carriers to bring up trade goods to Tinto and Bali, the soldiers and labourers on these Stations had to go to receive their pay in goods at Mundame, a circumstance which encouraged the Bali to roam about in the forest country. Von Stetten thought it was not unlikely that, given leave to roam freely, the coastward movement of the Bali could assume greater proportions; he puts forward for consideration the suggestion that an attempt should be made to guide this immigration into more orderly channels; this, given their possession of arms, would only be attainable by very adroit handling of the situation.

Notes

1. Max von Stetten (b. 1860), properly Max Freiherr von Stetten-Buchenbach, joined a Bavarian cavalry regiment in 1879. He went to Cameroon in 1891 as a volunteer in the service of the German Foreign Office and took part in the campaigns against the Abo and the Bakweri of Buea.

    In 1893 (by which time he was denominated *Rittmeister*) he led an expedition from Balinga Station via the hitherto unknown Tikar country and Ngaundere to Yola. In 1894 he briefly became Commander of the Protectorate Force. After his retirement in 1896 he became the director of the Royal Stud Farm. His later mentions of the Bali (known to the Tikar) are not unfriendly, and in 1894 he recommended that the Tinto Station be re-opened. The report of his expedition to Yola is to be found in *DKB 6,* 1895.

2. Von Stetten's ethnography is a simplistic trade geography which takes no account, for example, of the long discontinuous Bafo enclave that separates the main settlement areas of the north and south Bakundu. In this respect, it adds nothing to Zintgraff's reports in *Nord-Kamerun.* Esser's route map leaves a number of settlements unnamed, some designated as '*Niggerdorf*', i.e. *ninga* or slave-settlements. His route to Bali passed through Mokonge, Mamban (Mambanda), Baduma (a Balong village), Ambulo, Bulo Nguti, the settlements of Ndoi and Bakundu Nando, Bakundu Bakonge, Kombone (a Bafo village where a river was crossed). Then Nonge Maliwa and Kamfo (Nkwenfor), nowadays claimed for Bakossi migrants. To the west of these is shown Dikumi (Bafo) on a hill-top. Then comes Kokobuma (Bafo), Talangi (Talangaye), Mayimen (Manyemen), Bakun, Tobia, the last clearly a misprint for Fobia, the Guti (Nguti) villages, the Banyang villages of Difang Nonge, Tinto, Fotabe II, then a long haul to Sabi (Sabe's) and from there to the highlands via Banti, Aschu and Babessong to Bali. To the west

of the last leg of the route we read 'Passini, destroyed', evidently another misprint. On the return journey, as we know, Esser made a detour to Tayo, a Banyang place, and apparently made another between Tinto and the 'Guti villages', via Fotabe I, unmentioned in his text. (A Fotabe, so called, was a rival to Zintgraff's Fo Difang or Defang). Not much later Conrau could designate both Mundame and 'Mokonge' as Balong villages, Kumba, Ikiliwindi and Batom as inhabited by Bafo as are Mambanda and Kokobuma, while Bakuni is placed among the Banyang (see Conrau 1894b, and 1898).

That inveterate traveller F. W. H. Migeod (1925) trekked to Nguti in 1922, seeking out 'tribes'. Passing through Ikiliwindi (Bafo), he reached 'Bolo', then still a Bafo place, after 12.5 miles (it has since been colonised by Mbonge and appears on maps as Bolo Muboka). He then passed through a mixed Bakundu and Bafo area which included Ndo (Ndoi) and the Bakundu village of Konye (Esser's Bakundu Bakonge) which had recently moved a short distance. Then he passed Dikume (Bafo), Kombone (Bafo), Nonge Madiba (Maliwa) a 'Basossi' village he avers, and Talangayi [sic] (Balong) which he leaves undescribed. Then he reached Manyemen, a completely detached Banyang group who, he writes, called themselves Bachui, as Thorbecke (1911) and others had recorded, to reach Nguti. By then it had become a rather mixed settlement around a Government school. Contemporary sources identify it as historically a settlement of the Mbo Basossi cluster, the main town of the Miyengge group. For the Difang (Defang) mentioned by von Stetten see Zintgraff (1895) chapters 4 and 5 and Ruel (1969), 9, 13, 27. The Bakuni or Bachui of early travellers presented a puzzle to later-comers since, to quote Ruel (1969, 10) they were 'wary of the local government implications of claiming kinship with the Banyang' and said they had come from Manenguba, calling themselves 'Upper Balong'. The linguists, who call them Bakoni, have confirmed their Banyang origin, which was known to Zintgraff (1895, 132). See Map 2 for their approximate present location.

The communities lying close, most a little to the west, of Esser's presumed route, who kindly gave their time to one of us (U. R.), had no recollection of Esser's passage. Their memories concerned matters of greater importance to themselves as recognised administrative units.

3. See G. Böckner (1892, 1893) and Chilver (1967a). The situation blamed solely on the Bali was a good deal more complicated than the report writer was aware of, it seems. The abandonment of sites along well-travelled military and caravan routes and the re-siting of villages in the face of frequent large demands on their services and the food supply was a fairly widespread phenomenon at this period and later, quite apart from the plundering of farms and stores by armed parties. For another possible cross-current – the presence of youthful bands imitating the German forces – see J-P. Warnier (1996) in *African Crossroads,* 115–123. It may have already begun to make its appearance.
4. There were no distinct slave settlements on the forest pattern in Bali-Nyonga: slaves resided in their masters' compounds or might be sent to live and work on outlying farm settlements to clear bush, build huts and fences, tend livestock, harvest and guard crops, and defend women cultivators against kidnappers.

12

A Complication: the Entry of the *Gesellschaft Nordwest Kamerun*, 1901–1903: Esser's Correspondence

The privileged position of the WAPV as, in some sense, the middlemen between Bali-Nyonga, and a distant Empire and ally whose power could be exercised in Bali's interest, was challenged by the entry of the GNK.

A huge concession area in the northern and western hinterland, which included the Cross River basin, the Grassfields, the trans-Mbam area, much of Adamawa and more vaguely defined areas 100 km from the coastline, were initially accorded to the company in September 1899 on condition that land would be made available for government-approved purposes free of charge, and that freedom to trade was to be respected, other than in the products of the so-called Crown or ownerless land. The concessionaires were under the obligation to explore the area, to make roads and to establish trading, agricultural and mining enterprises. The company was empowered, with the Governor's consent, to acquire so-called 'ownerless land' and, for twenty years, granted priority in the purchase of land from 'the natives'.[1] The founders of the concession company included a princely Silesian magnate, Prince Hohenlohe-Öhringen, the Düren industrialist Max Schöller, and the landowner and factory-owner Max Hiller who was, until 1902, also on the board of the WAPV, which naturally felt its privileged position in Bali threatened by the GNK. Esser addressed both the Governor and the Foreign Office on its behalf in December 1899 (BAB, R1001/RKA 3499, 133–136) describing WAPV activities in Bali which, apart from labour recruitment in accordance with the conditions agreed with the Bali ruler, had involved the further development of the land granted for its Station, regular and often long visits by its recruiters (Steinhausen in 1897, Netz in 1898, Steinhausen and Bornmüller in 1899, Kionka in 1900), trade in 'cloth, beads, old uniforms etc. against ivory, palmwine, local produce and curios', as well as wild rubber and ivory brought down to the coast. The colonial authorities were asked to protect the WAPV's land and trading rights against the GNK. In the Governor's absence his deputy replied that existing trading activities would be protected but that he was unable to deal with the land matter since no written record of the agreement had been submitted for approval: a Land Register now existed for the whole protectorate. To judge from the WAPV annual report for 1899, Esser had not given up hope of an amicable compromise with the GNK, which was not forthcoming. Further correspondence (in the Federal Archives, Berlin-Lichterfelde) with the Colonial Section of the Foreign Office ensued which is summarized in a recapitulatory letter from Esser dated 18 December 1901 (BAB, R1001/RKA 3501, 10–14) to both the Governor and Foreign Office. A translation of the main text follows:

> Dr Zintgraff, Hoesch and Dr Esser undertook an expedition to Bali in May 1896, at the cost of some 30,000 Marks to conclude a treaty with Zintgraff's friend Garega, its Chief, for the provision of labour in connection with the completion of the plan to develop plantations in Victoria and maintain a permanent Station in Bali.
>
> This expedition was fully discussed with the then Director of Colonial Affairs, Dr Kayser, at the end of 1895 and in the early months of 1896, and also with Governor von Puttkamer, who was in Berlin at the time. The relevant correspondence from Dr Zintgraff must be in the contemporary files and will indicate that both the Director and the Governor had assured these gentlemen that they would sanction any arrangements made with Garega, and that they were delighted that the expedition was to be undertaken.
>
> The Imperial administration equipped the expedition with 25 rifles and generously put the steamer *Nachtigal* at its disposal.
>
> After their return to the coast at the beginning of August 1896 Dr Esser and Hoesch informed the then Chancellor of the colony, Dr Seitz, of the successful outcome of the expedition and Dr Esser also reported this to the Director, Dr Kayser.
>
> As to the entry of the Bali land in the Land Register Dr Seitz concluded that this could await the return of Governor von Puttkamer since no Register existed yet for Bali. The land acquisition in Victoria was entered into the provisional Land Register by Dr Seitz himself.
>
> On a number of occasions Governor von Puttkamer had stated that he considered the arrangements with Garega and the stationing of a European company representative in Bali was well-conceived and conducive to the solution of the labour problem.
>
> Dr Zintgraff stayed up in Bali, after Dr Esser and Hoesch had left on the 21st

June, until the end of October and, at a relatively substantial cost, planted the 100 hectares acquired with kola, rubber and vegetables, setting aside a substantial part of the area for grazing. Please note that the enclosed extracts from Zintgraff's diary deal with his activities in connection with the cultivation of this land. And later, after Zintgraff had returned to the coast, first Steinhausen, then Bornmüller, later Weissenborn, were sent up to Bali – indeed, a company employee has been to Bali right up to the beginning of this year.

In our area, we have built 5 houses, cattle-stalls and so on, and 20 hectares are under cultivation. As soon as the statute of the GNK was published we turned, on 10 December 1899, to the Imperial Governor and to Your Excellency (at the Foreign Office Colonial Section) and drew attention to the fact that we had occupied 100 hectares of land and asked for the protection of our rights.

At your Excellency's request we later also presented a statement made by Garega concerning the transfer of land to Zintgraff, Hoesch and Esser on our behalf on 6 June, the validity of which was confirmed before an Imperial Judge on 23 July, 1900.

By this time Your Excellency had, according to your letter of 8th April, 1901, made inquiries of the Governor; during this period the company's representative was in Victoria and further established that our land-rights in Bali were incontrovertibly within the law and that our claim was further fortified by long occupation and cultivation. Yet the Foreign Office repeated, this August, that the GNK possessed all the land in Bali, without protecting our rights.

However since, in the Ordinance concerned, the issue is merely that ownerless land should fall to the GNK, we repeat yet again that the land made over to us by Garega has been kept in cultivation since June 1896.

The GNK in their letter to us of 30 December, 1899, have already stated that they would adopt 'a thoroughly negative attitude' towards our claims. In these circumstances we are no longer in a position to negotiate with them and must therefore ask Your Excellency once again to award us the land on the basis of Garega's deed and Zintgraff's diaries.

We further observe that when our Company was established it took over all the rights acquired by Messrs Zintgraff, Hoesch and Esser for about 50,000 Mks and that these rights were adjudged to be theirs at all events by Director Kayser as well as by Governor von Puttkamer, at least orally if not in writing.

Governor von Puttkamer would certainly be able to declare upon oath that he had informed the representative of our Company on several occasions that our rights in Bali were legally recognized. It follows that our land acquisition was orally approved by the Governor and the Foreign Office before the concession to the GNK was authorized.

Were Your Excellency not to agree with this exposition of our claim we would, to our regret, be obliged to bring the evidence for its approval before the courts. We would then not only seek repayment of the 50,000 Mks. already spent, but also seek to be indemnified for the losses we face occasioned by the removal of our influence in Bali and the resulting shortage of labourers in future years.

[sgd. Esser]

The enclosures consisted of the following:

(1) Relevant extracts from notes made by Hoesch and Esser, establishing that, in return for the presents they had brought, they

took over land given to them by Garega for cultivation and had concluded a labour agreement with him. There follow extracts from Zintgraff's diary describing what he did after their departure, e.g. clearing the station area. The entries ran from 21 July 1896 to 22 August 1896.
(2) The Agreement with Garega drawn up by Kionka, and the minutes of the Agreement.

This is soon followed on the same file (at 16) by a further letter to the Foreign Office dated 23 December, 1901, from Esser on behalf of the WAPV, as follows:

> With reference to our letter of the 18th of this month concerning our property in Bali our board is now in the position – supposing, that is, that written approval for our land-acquisition in Bali may thus be made easier for Your Excellency – to propose as follows:
>
> We ask that if this approval is to be recast we be permitted to make a 100–year leasehold agreement with the present Chief, Fon-Yong, whereby he leases to us the land made over to us by his father. Since this land belongs to the village area of Bali and is, in no sense, ownerless land it cannot, in our view, ever become GNK property on legal grounds because this company only has pre-emptive rights of purchase and not any pre-emptive leasehold rights over land in native ownership. Moreover we ask Your Excellency to bear in mind that we do not value the 100 hectares equally highly but attach the greatest importance to our buildings and cultivated land (for kola and potatoes and grazing meadows) which perhaps occupy some 25 hectares.
>
> We ask for your sympathetic consideration of our proposals and suggestions and reiterate that our relations with Bali and our properties there are of the greatest importance to our plantations and have been supported in multiple ways, though not in documents, by Your Excellency's administration which has both approved of them and made them effective. We hope for a favourable solution and ask you to note that our representative is ready to discuss matters in person at any time.

The disturbance and subsequent heavy-handed punitive expeditions that followed the deaths of Lt. Queiss and the trader-recruiter Conrau[2] slowed the deployment of the GNK in the Cross River and Bali areas. By September 1900 it had established a post at Nssakpe, near the Cross River rapids in the Ejagham country. An enlarged expedition led by the GNK's local director, von Ramsay, reached it in early October and soon after set out to explore to the north and north east of it in a wide half-circle through often hostile areas. It reached Bali from the west in mid-November and was greeted with enthusiasm by the ageing Galega, to whom Ramsay promised 'a big factory'.[3] Ramsay's description of the festivities (in Hans Zache's collection of colonial reminiscences, 1925) resembles Esser's: Ramsay, too, was impressed by Galega's kingly bearing. The company's reports

(1899–1903) stress the potential contribution of the 'Bali lands', the S.W. Grassfields, to labour supply. The close initial cooperation between the GNK and the newly-formed (1895) Protectorate forces resulted, in the autumn of 1901, in a joint visit to Bali by Captain Hans Glauning, then head of the Nssakpe military post, in the company of the GNK's representative, then Graf Pückler-Limpurg, following an earlier probe by a small force under Lt. Strümpell from Tinto to clear the way for a major expeditionary force. By the end of 1901 Pavel's large force had arrived in Bali, which was in readiness for it and went on to 'punish' Bandeng (Mankon) and Bafut and capture and exact penal labour before marching on to Adamawa by way of the Ndop Plain and Nso'. The establishment of a military station at Bamenda followed in 1902. After 1902 an official licensing system for private recruiters protected the WAPV's access to the Grassfields, but there was now to be strong competition between the demands of the military station and other users.

Little of the rougher side of the GNK's operations is to be found in Ramsay's publications. It was not locally popular for reasons that emerge from a report sent by one of its managers to the directors, explaining why the Chief of Bascho refused to allow a company agent to settle there:

> He informed Herr Kuester (the GNK's agent) that, while he and his people were pleased that a store had been built in their chief village, they could not put up with the permanent presence of a European, for even if only one was there at the start, more and more would come and finally the soldiers would arrive – and that spelt endless trouble. (Diehl to GNK Directors, 23 April 1903, on BAB, R1001/RKA, 3469, 168–170).

The atrocities of hungry and ill-supervised soldiers on caravan routes, escorting supplies or penal labourers, were reported by the Basel Missionaries whose complaints began to receive support in the *Reichstag*. (See, for example, letters forwarded by Mission Inspector Lutz, August 1905, in BAB, R1001/RKA 4433, 5–13). Apart from this the forcible methods the concession company often used in recruiting carriers, compelling road-work and claiming free supplies of wild rubber added to an unpopularity aroused by its interference with local trade.[4] The revolt, of 1904 in which Boki, Anyang, Keaka-Ejagham and lower Banyang combined to loot the GNK factories at Bascho. Badje, Abokum and Mamfe, the Government Station at Ossidinge and the customs post at Nsanakang, and in which five GNK traders and the District Officer Graf Pückler lost their lives, was partly a reaction to such methods, which included the supply of recalcitrants as labour to the coastal plantations. Endless recriminations followed it, with the GNK seeking damages for its losses and shifting the blame for the rising on the provocative actions of the soldiery.

Notes

1. This in fact monopolized the main articles of trade, wild rubber and ivory in particular, while the lack of competition lowered prices to local producers. For a detailed account of the opposition to the GNK see Jolanda Ballhaus in Stoecker (ed.), Vol. 2, 1968, 130–173. Its area of operations was reduced and in 1910, its privileges were revoked. It failed in the task it was set, its attempts to drive out competitors had little success despite an agreed take-over of the Hamburg firms' trading posts and gave rise to a diplomatic contretemps with the British over the rights of the old-established firm of John Holt's in the area, under the Congo Basin Treaties. Some of the claims of its agencies and subsidiaries had little behind them. It proved to be a disastrous investment for its promoters, unlike its southern counterpart, the GSK. For a well referenced case-study see Rohde (1997), 87–100.

2. The punitive expedition against the Ekoi (Ekwe-Ejagham) of 1900 was led by Hauptmann von Besser. Lt. Queiss, accompanied by soldiers and 120 carriers, had been killed in 1899 in a battle near Otu by Ekoi resisting the German advance. Conrau had been sent to help Queiss and inquired into the circumstances of his death. Conrau, who had earlier recruited 50 Bangwa labourers for the WAPV, returned in November 1899 to pick up his belongings which he had left in the care of the Fon of Fontem in Bangwaland. Meanwhile, it seems, rumours had reached the Bangwa that labourers sent to the coast had died there. Confused circumstances surrounded Conrau's death in 1899 by his own hand, fearing capture by the Bangwa, who had held him responsible for the non-return of labourers he had recruited. The successive punitive campaigns which followed his supposed murder are described by Chilver in the *Journal of the Historical Society of Nigeria* 4 (1), 1967, with reference to contemporary official accounts, as a sequel to the oral Bangwa account recuperated by Elisabeth Dunstan in an earlier issue.

3. See also Ramsay's report of December 1900, which is cited by Rüger in Stoecker, Vol. 1, 1960, 202 (BAB, R1001/RKA 3469, 3).

4. For the view of a Bali employee of one of its agencies, involved in the cattle-trade, see Max Fohtung's reminiscences in *Paideuma* 38, 1992, 224–226. The agent in question was still almost laughingly remembered in Nso' in the late 1950s, as something of an irascible figure of fun, who beat his carriers with a furled umbrella, and was nicknamed 'Masta Tisong' (Master Station).

13

A Parliamentary Visitation: Johannes Semler's *Togo und Kamerun: Eindrücke und Momentaufnahmen von einem deutschen Abgeordneten*, Leipzig, 1905

Dr Johannes Semler (b. 1858 in Hamburg), D. jur., was, according to the *Deutsches Kolonial-Lexikon,* a lawyer active in Hamburg civic affairs, who, in 1900, became a member of the *Reichstag* in the National Liberal interest. He played a prominent part in debates concerning colonial budgets and legal issues. In 1905, not long after the alarming Anyang revolt had been finally put down, he visited Togo and Cameroon as a member of a parliamentary study group and immediately published his impressions, illustrated by 37 of his own photographs, anonymously. These 'impressions' included conversations with plantation managers and officials. The party was taken around the WAPV plantations and was given a lecture by their manager, W. van der Loo, on the processes and demands of cocoa production. They sought for information on the condition of the several thousand labourers on the plantations, both free and forced. Both Bali and Wute are mentioned (29–30). Dr Semler's remarks reflect some of the concerns of the trading and manufacturing interests he represented. The page numbers of Semler's book are shown in brackets.

I do not myself think that it will be possible to recruit permanent voluntary workers from the hinterland since they, too, suffer considerably in the coastal climate; moreover it would be undesirable to depopulate the inland districts. In this matter, the Station commanders seem to be right in being unwilling to give this their support. Insofar as forced labourers are concerned we should like to see their numbers reduced to a minimum. These labourers are those Blacks who must be presented by chiefs who have shown disobedience – either such or else war-prisoners. It is understandable – and this was confirmed by officers of the Protectorate forces – that under such circumstances chiefs present men of lesser value to them, mostly slaves. We later had an opportunity, at another plantation near Man o'War Bay, to come upon such a Black man. We were making for its so-called hospital. On the way there there came, or rather crawled, towards us the pitiable figure of a Negro, hollow-cheeked, grey, just skin and bone. With hand signs to his mouth he signalled his hunger and we immediately called over the Chief Medical Officer who was accompanying us – he was incidentally, the principal report writer on medical affairs in Cameroon. The latter was taken aback, and had some rusks brought which the poor fellow fell upon ravenously. In the hospital we came upon three other poor creatures that hardly looked like human beings, crouching in the hot ashes of their fire, but they crawled away on our arrival. Our horror must have shown on our faces. We had come here unannounced and the thought that was uppermost in our minds was that we had had a glimpse of misery which might have been removed from our sight if our visit had been expected. The farm manager must have sensed this because, when the Medical Officer, Dr Ziemann, sought him out, he immediately offered to summon all the labourers within reach so that we could make an assessment of the nutritional state of his people. Then the planter's evident agitation burst forth: we heard what was on his mind. The men we had seen were Bakum[1] war prisoners, sent to him as labourers after the last rising, ailing creatures he had been obliged to feed up, but he had not managed to in these four cases. His plantation was not a sanatorium for Black invalids. Such people should not be sent to the coast – and so on and so forth. When he then conducted us through his plantation the experience we had just had was still so vivid that our attention to this well-laid-out plantation was perfunctory. As we wandered around we found that some hundreds of labourers had been rapidly summoned, and, fortunately, they did not give the appearance of being badly looked after. Their stomachs were swelled with the staple Negro food, the banana or plantain. So it seemed that what our medical friend had meanwhile told us, namely that this plantation management did not have a bad reputation at all, and that those shocking appearances of misery could well be due to special circumstances, might be true. We looked into the circumstances and were told that uneaten food had been found under the men's blankets. It remained uncertain whether the sick men were suffering from dysentery, or desperately home-sick, and whether their behaviour was wrongly casting aspersions on the plantation manager . . . Since we were not an inquiry commission we purposely took no further steps in this matter.

But we had all come to the same conclusion, that the transport of prisoners of war to the coast in the present manner should be brought to an end as soon as possible, not because of *our* views but because of those of more knowledgeable persons and that, in particular, inland Stations should not send sick wretches on this journey when they were unfit to sustain its hardships. That this does come about we were again able to confirm while on a railway trip around the WAPV plantation ... Could one not, as seems possible in Togo and as we observed in Yabassi and in the English colonies, hold prisoners for long enough at the Station – they are chained together to prevent escape – either to improve their health, or otherwise, to return them to their chiefs as unusable?

As to the rest of our tour it is happier to report that sick labourers are well looked after once on the plantations themselves. At Bibundi plantation as well as Idenau Sanje we saw hospital quarters with pharmacies, Black 'doctors' and White dressers which were favourably regarded by everybody. It is, of course, in the interest of the big plantations to make such provision so that the common complaints affecting hands and feet can be dealt with quickly and the duration of dysentery and malaria reduced as much as possible. I believe that the resistance of the inland tribes to plantation work, perhaps only among the Wute as the Fulani and Hausa will never do such work, may diminish, given time. It would be nice if the labour question could be permanently resolved for the present number of established European plantations. However, I remain of the opinion, which is not merely based on the impressions gained in a short visit, that the future of Cameroon, and especially that of the fertile coastal strip, does not lie in large-scale European plantations, but in African ones.

One such was shown us by the manager of Esser's plantations; this stretched up from the coast to Esser's property. Our mentor spoke freely of its good points. The Black owner employed his own people and slaves and it produced an up-to-date and tradeable crop even if the plantings seemed less orderly than Esser's. Nevertheless, added the manager, this was an exception to the rule. The surrounding Bakweri were on the whole a lazy lot. Just look at their reserves, he said, as we went through them. It was only where the plantation railway had made clearings for them that their slovenly farms were to be seen, and anyway they were too lazy to make anything of the reserves granted them. One detected that the native reservations and the activities of the 'Protector of the Natives' (a statutory member of the Land Commission) were seen as hindrances by this manager of a great plantation, who thought that native land needs had been greatly exaggerated. (31–36)

Dr Semler dodges the issue: it was too complicated, he says, to tackle on the basis of a short visit. But millions of Marks, to his knowledge, had been invested in the plantations which had not yet been able to show any substantial profits. All the same the Blacks should have security of tenure in their settlement areas. On the other hand he accepts the need for compulsion to instil regular habits of work and impart useful training, on the analogy of compulsory military service in Germany. This could best be done in a man's own district. Funds, he suggests, should be found to set up experimental plantations elsewhere than just in Victoria, and natives obliged to work on them for a year to gain appropriate agricultural experience. Such a plan would have to be introduced cautiously to avoid resistance. Semler repeats his view that the future of Cameroon lies in 'African plantations' for cash crops.[2] Had such a policy been started 20 years ago the banks of the Sanaga, the Wuri

Figure 13.1: Penal labourers (J. Semler, 1905)

and the Mungo would have been a forest of oil palms and fulfilled Woermann's wishes, expressed when he said: 'Get me palm oil and palm kernels from the lands which can still produce huge quantities of them, for they are, in the last analysis, as important as cotton and other tropical products which call for elaborate planting.'

Semler, as the Governor's guest, is cautious about the Basel Mission's advocacy of 'native grievances' and gives no details of them. On the Bakweri lands questions he did not think that, during their party's interview with the Mission representative, the Missionary had tackled the larger economic issues involved: his prejudice against the big plantations was evident (53–54).

Semler has more to say of interest on colonial district administration – few officers stay long enough in one post to understand the

Figure 13.2: The Schloss at Buea (J. Semler, 1905)

surrounding peoples – and on the need for prior training and a career service.[3] He doubts whether its militarization is a good thing and cannot resist a smile at some Prussian mannerisms, exemplified by the wearing of jackboots and monocles in the bush. Railway prospects are discussed. He had a long interview with Manga Bell and had suggested to him that his son Rudolph should be sent to Germany to be trained as an engineer rather than to Lagos to be trained as a lawyer. The book includes three illustrations of African women, two evidently posed, but they are otherwise invisible in the text. After expressing his own preference for a regime of universal free trade rather than the acquisition of colonies he concludes:

> ... But once we had acquired colonies, and meanwhile installed a system of protective tariffs in the agrarian interest, we also needed to set out to encourage the homeward export of those tropical products our industries cannot do without, and which, otherwise, would have to be imported from other tropical lands or the United States. Thus our colonial policy is, to a considerable degree, a sequel to our agrarian policy. It follows that we should do more for our colonies, especially Cameroon, than in the past, and not merely act in response to the pressure of native risings. We should be ready to reconsider the principles which have so far been employed in their administration and we should not shrink from sacrifice on behalf of their good government on the one hand, and the opening-up of the colony on the other.[4] (*101*)

Semler, as might be expected of a member of a government-supporting alliance of parties, is in general reassuring and tactful and not disposed, despite his uneasiness about penal labour, to make accusations by name against officials, mostly represented as good fellows. This was the self-appointed task of Matthias Erzberger, leader of the left democratic wing of the Centre Party, in *Reichstag* debates, books and pamphlets.[5] His targets included Governor von Puttkamer, and Hauptmann von Besser.

Notes

1. Probably Mbakem, an Anyang village north of the Cross (Manyu) River. For a general description of the situation following the suppression of the protracted revolt see Mansfeld (1908) 18–21, and Moisel (1904) for a map. Mr Daniel Awu, an elder of Agborkem, an Ejagham village, interviewed in February, 1999, told U.Röschenthaler that one of its chiefs had been involved in the killing of a GNK trader and had fled from the German forces to his Anyang friends at Mbakem. Finally brought back as a captive to his village he was hanged with his sons. Diehl mentions the opening of a store at 'M'Bakum' in 1903 in a letter to the GNK management (BAB, R1001/RKA 3469, p. 170).

2. This appears to echo some features of a memorandum by Zintgraff of 1892, in which he proposed that grants and technical advice should be given to Africans of standing to enable them to start family plantations devoted to commercial crops; the Government was to have a lien on their eventual production at fixed prices. This proposal, published in *DKB* 3, was accompanied by a deflating commentary by the Colonial Section of the Foreign Office, pointing out the huge staff costs of managing the scheme. For an extended paraphrase see P.N. Nkwi (1989), *The German Presence in the Western Grassfields*. Leiden, 23–33.

3. See Hausen (1970) 115–140, for the metropolitan history of colonial service training and for a close analysis of the career prospects and attitudes of the military and civil administration in Cameroon.

4. Semler's polite rhetoric is far from clear on the terms of the debate between plantation and trading interests, which came into the open after von Puttkamer's recall, and dismissal in 1907. Some of the positions taken on the 'native peasant farming' issue are described by Wirz (1972) 208 ff., and van Slageren (1972) 72–73.

5. For a useful summary in English see Klaus Epstein's 'Matthias Erzberger and the German Colonial Scandals', in *English Historical Review*, 74, 1959, 639–663. Von Besser's careless brutalities are described in some detail in Erzberger's *Die Kolonial-Bilanz*, 1906. For a view of his wider political and party background see Epstein's *Matthias Erzberger and the Dilemma of German Democracy*, Princeton, 1959, 38–95.

14

A Soldier's View of the Tasks of the Bamenda Military Station in 1908: Hptm. Menzel

Appended is a translation of extracts from the half-yearly report, 1 April to 30 September 1908, to the Imperial Government in Buea. It comes from a damaged file, *Verwaltungsangelegenheiten: Bezirk Bamenda,* IC51^n, found and repaired by Edwin and Shirley Ardener. A photocopy is to be found in the Rhodes House (Bodleian) Library in the MSS Africana Collection (MSS. Afr. S. 1529). The original file has been transferred to the Fonds Allemand in Yaounde from the Buea Archives after Eldridge Mohammadou compiled a catalogue[1] in which it did not figure. The writer, Lt., later Captain, Menzel, took over the division after the death of Captain Glauning,[2] who had taken over the Division after a serious uprising, the Anyang revolt of 1904 nearby, and had rewarded the Bali ruler for his loyalty at a time when other groups had reasserted themselves and refused obedience to the Station: the sphere of Bali agency was, temporarily, greatly increased and Bali auxiliary forces assisted the garrison in punitive actions.

> The programme for the pacification of Bamenda was: temporary subjection of the whole area, to be followed by the spread of several detached posts throughout it. However, the requirements of labourers and carriers laid on the Station the task of taking systematic levies from individual tribes. This could only be achieved by the thorough subjection of chiefs and by ascertaining the ability of the tribes to provide the carriers and labourers required. As the Station was forced by the situation on the British border to subject further tribes, the complement remaining

at the Station was too weak to enforce obedience to its orders and it met with difficulties in meeting demands for labourers and carriers.

As a consequence a whole string of newly-subjected tribes was put under the authority of the Bali chief and, in the course of time, gave him a certain influence over the tribes to the south as far as Bandjoun. Because of its continuous expeditions, the Station was nearly always undermanned by Europeans and soldiers. So, in order to simplify the machinery of command [*Befehlsapparat*] intermediaries were used; for example in Fumban [Bamum] before the establishment of the experimental [agricultural] station, the European factory staff and later the [Basel] Mission. In Widekum orders were transmitted to the Chief by a coloured factory manager.

It has become evident that these intermediaries are using this position for their purposes, too, and think they can gain or increase their own influence.[3] It follows from this that closer contact between the Station and the various chiefs, individually, has become impossible, and this has been harmful to the influence and prestige of the Station.

During the last long absence of the Station's forces and because of the death of the *Stationschef* above all, this situation has become untenable despite the work and effort of his deputy.

With the end of the Munschi Expedition[4] [against peoples on the Division's north and western borderlands] the area has, more or less, been pacified. Now the actual work in the Division's area must start. As Captain Glauning was the only person with accurate knowledge of the Division his death could not have come at a more inopportune moment. To this must be added the fact that each new *Stationschef* has to reckon with passive resistance from various Chiefs, for every African Chief will first try to contract out of the requirements laid upon him by a new *Chef*. This applies equally to the Europeans in the Division.

Therefore, the Station has had to make a new start to establish direct contact with the Chiefs and to remove the intermediaries. The pacification of the South [this refers to a campaign in the S.E. Bamileke region] and the successful outcome of this task as regards the chiefs on the banks of the Nun, and those of Kom and Nso' shows that the Station has not met with special difficulties thereby [i.e. in doing without intermediaries]. This task is likely to be much more difficult in Bali and Bamum where the Station has to reckon with the influence of Europeans who are not responsible to the Government.

There follows a passage about Bamum, concerning the permission given to its ruler Mfon Njoya to maintain an armed bodyguard and its possession of some old breech-loaders, which had excited European gossip, as well as the reasons why Bamum auxiliaries had been used against Nso' in 1906. Menzel continues:

> Conditions in Bali are less favourable. The former enemies of the Chief have been placed under his jurisdiction. According to the conceptions of the Black Africans these areas are to be exploited for his own ends and those of his Balis. It is my firm opinion that only fear of the Station prevents these badly oppressed areas from freeing themselves from the Bali.

The report then refers to an exchange of fire between the 'Bali soldiers', the armed force permitted to the Chief of Bali and used in

Figure 14.1: The proclamation of Fonyonga II as Paramount Chief on 15 June, 1905, in the presence of Captain Glauning, Lt. von Puttlitz, the Basel Missionaries Ernst and Keller, soldiers and delegations from 47 other Chiefdoms: photograph by Missionary Goering. (Reproduced by kind permission of the Basel Mission: photo archive No. K. 776)

early years as auxiliaries by the Station, and Babessong, earlier armed by Zintgraff.

> To prevent such incidents from recurring the Station was obliged to prohibit the use of the Bali force for such purposes. Obviously, the Station must ensure that the Chiefs placed under his [the Bali ruler's] jurisdiction obey him. However, in order to end the situation the Station will be forced to dictate to the Chief of Bali how many labourers and carriers he may take from each of the Chiefs subjected to him. To estimate their numbers a European is now visiting the Bali sub-Chiefs.[5]
>
> In my personal opinion to deal with the Bali question is the most important matter and the only one that will cause difficulties.

The report now turns to other matters, including the destructive exploitation of wild rubber and the increasing popularity of the kola trade. In subsequent reports the *Stationschef*, Menzel, maintained his critical attitude to Bali, which provoked some surprise in the Governor's office in Buea which minuted in mid-February, 1909:

> In the past, the Bali Chief was generally the favourite. It is evident that he did a lot for the Protectorate during the Zintgraff expeditions. Government expeditions have always engaged his extensive support and, in the past, Lt. Colonel Pavel sent him women and children taken during the Bafut war in recognition of his

Figure 14.2: Bali Soldiers (*Basoge*) in a motley of German uniforms: c. 1907–8. (Ankermann Collection, No. 315 D – Reproduced by courtesy of the Museum für Völkerkunde, Berlin)

services.[6] Without Bali labourers the development of our plantations, especially those of the WAPV, and its caravans, would have been impossible. A number of the large ivory tusks which Governor von Puttkamer presented to His Majesty were donated by Bali . . . Bamum seems to know better how to make itself agreeable. Nevertheless, one should not forget the former services rendered by Bali, which were always mentioned by Captain Glauning in special terms.

Menzel went on leave some months after this despatch but took over again from his deputy Raven early in 1910. Both continued to express uneasiness over the extension of the Bali paramountcy as a result of punitive expeditions between 1905 and 1907 and over Bali demands for support as involving an expensive use of force in a Division which was, as yet, far from 'settled', according to their lights. Menzel's needs for carriers, road labour and station maintenance could only be filled where the Station's power could be felt, or the threat of it feared, '4 or 5 days' march from the Station' and where labour in lieu of the head-tax (extended to Bamenda in 1909) could be exacted from obedient Chiefs. In 1911, he was succeeded by Adametz, to whose views we shall next turn.

Notes

1. E. Mohammadou, *Catalogue des Archives Coloniales Allemands du Cameroun: Le Service des Archives Nationales de Yaounde,* 1978: Institute for the Study of Languages and Cultures of Asia and Africa, Tokyo. The Bamenda papers are now registered in the FA 1/110 Series, Yaounde Archives.

2. Hans Glauning (b. 1868, killed in action 5 March 1908) became a Lieutenant in a Saxon regiment in 1889; he transferred in 1894 to the Protectorate Forces in German East Africa where his skills as a survey officer were in demand. He joined the Cameroon force in 1900 and commanded the first military station in the Cross River area. In 1902, he accompanied Pavel's Chad expedition as a topographer and he was a member of the Chad-Yola Boundary Commission. After his arrival in Bamenda most of the region was held to obedience by sharp punitive patrols and campaigns in support of obedient chiefs as well as by well-armed explorations. These called forth unusually informative despatches and good maps. He both collected and commissioned Grassfields arts and crafts and left his collection to the Berlin Ethnographic Museum *(Museum für Völkerkunde).*

3. From 1903 onwards Missionary Ernst had established close relations with Fonyonga and took upon himself, almost to the exclusion of other Basel missionaries, the task of advising him: Ernst was largely responsible for converting the heroic tale of the great raid of the Chamba leader Gawolbe into an 'empire' described in quasi-feudal terms, to which Fonyonga might be seen as a legitimate heir. Hutter's earlier assessment of the political situation of the Bali-Nyonga in numerous articles and in his book was, it seems, overshadowed. How far Glauning was influenced by Ernst's views rather than by Bali's services is uncertain but he was evidently more ready to use the Mission's diplomacy and on better terms with it than his predecessor von Knobloch, whose unruly soldiers were complained of. Menzel, before going on leave in May 1909, accused Ernst of interfering, in Bali's interest, in a dispute on the Bamileke border.

   For an excellent analysis of the shifting triangular relations between Bali, the Mission, and the Military Station see the Basel University dissertation by Andreas Merz (1997), *Die Politik Bali-Nyongas gegenüber der Basler Mission und der deutschen Kolonialmacht.*

4. A convenient summary of the so-called Munschi expedition can be found in P.N. Nkwi (1989), 64–72.

5. Menzel here anticipates the issue of more comprehensive labour regulations which placed private recruiters under the authority of Station and Post commanders and required them to present these officers with regular returns and reports of their activities. For an English summary see P.N. Nkwi (1989) 74–75.

6. The use of captured women for the gratification of regular or auxiliary troops was not unusual; some were sold on. In response to a British inquiry concerning a woman who had passed through at least four hands, Mfon Fonyonga II of Bali-Nyonga wrote on 4 December 1935:

> The woman was brought here during Germans [sic] when things were going on roughly. They were captives of the Germans during the time of patrolling villages of this part and then given to German soldiers by German white men as a present for fighting hard against any village they went to. The soldiers in turn sold them to Bali. (File BNA 140, Bamenda Native Authority Registry).

Similar anecdotes could be collected during fieldwork from former German allies. It seems probable that Menzel personally deplored the practice since he went out of his way to trace and return captured women to Nso'.

15

Labour Supply: a Shift of Modalities, 1913: Hptm. Adametz

The paper and extracts reproduced here had been removed from the German file IC51^n by the incoming British civil administrators and translated with the help of German-speaking Cameroonians. They were found in Bamenda among papers dealing with the puzzlements of the Nigerian Political Service in those areas in which Bali-Nyonga had claimed suzerainty.[1] These include a number of reports from W.E. (later Sir William) Hunt, Divisional Officer, enclosing translations of German papers found in Buea.

Among these were two lists of 'sub-towns' of Bali, the German *Vasallendörfer*, one dating from June 1905 and another from May 1914. The make-up of the first list had been a gradual process starting in 1904 and added to in the course of punitive expeditions in which Bali auxiliaries had played a part.[2] A major event had been the public proclamation, in the presence of other Chiefs, by Captain Hans Glauning, of Bali paramountcy over 31 villages (*DKB* 16, 1905, 667–672) which, as might be expected, was variously interpreted by the different parties concerned, and altered over time. By early 1914, to judge from the last German annual report by von Sommerfeld, the 'vassals' had been reduced to eleven 'headmen'. Before that two Bamileke and two Moghamo chiefdoms had been lopped off the list. In December 1912 Fonyonga II had been publicly lectured by the Governor and told that he had shown himself to be:

> unable to distribute equitably among his vassals the services required of him by the Government, that is, tax-collection, carrier supply and labour supply.

For these reasons the taxes would in future be collected directly by Station Staff. All orders would be transmitted by Station Messengers and 'vassals' were to have the right of direct access to the Station and be allowed to keep a representative there. Meanwhile Fonyonga was to remain titular overlord of the villages assigned to him in his *Schutzbrief*, apart from those removed from the list. He would receive an annual allowance from tax-funds, to be agreed by the Station, representing the tax-commission (10% of the head-tax collected) that he had lost under the new system.

Other large changes were projected at the time – the division of the huge *Bamenda Bezirk* into two, a Residency in Bamum, an end to private recruiting groups and a larger role for the divisional administration in the estimation of labour supply possibilities. The Station itself would have its own heavy needs for carriers and road-makers, and with prospects for European settlement and developments in farming and animal husbandry in the offing, it might be possible for men to earn wages in their own district, it was thought.

In May 1913, Adametz replied, not without a touch of irritation, to a Government circular giving estimates of the labour needs of the plantations in the following terms:

> Your circular No. 777 has already been replied to . . . In our report we stated that the local authorities, as they are on the spot, should be authorised to decide where the labourers should be sent. Except for the Bali tribe and from peoples of the escarpment south of Bali, who have for many years supplied people to the WAPV, the Bamenda Division can only supply people for the Northern Railway. It would be more sensible to supply the Cameroon Plantations from their neighbouring districts and the well-populated Ossidinge division rather than from the high Grassfields. Only the southern part of the division is suited to such recruitment and the boundary might be fixed at the Banso [Nso'] and Bekom [Kom] tribes. These are hill people and known to be unfitted for and to suffer more in the coastal climate than other Grassfield people.
>
> In the southern portion [of the Division] we estimate, partly by means of actual counts and partly by estimation, that the number of men fit for work is 30,450. It is known that about a twentieth of these is free to work. The Division could supply about 1500 plantation labourers over the next two years. As carriers for traders, Missions, Government, tax collectors, and also for Kuti [the agricultural station] 3800 men are required. Together with the 1500, this gives 17Ω% of the number of men. As 80% of the men are needed to maintain the food supply, 1500 men are the very most that can be spared. I have not included Bamum, as it must be left out of account for the next five years – it has a strength of about 10,000 men – as having supplied 750 labourers (502 being men) to the Tobacco plantations for 3 years. They have also responded to a steady demand for carriers and supplied a regular force for Kuti experimental station. Moreover, a Residency is shortly to be established there, so a heavy levy cannot be imposed. The three Bamileke areas containing 13,520 men, which are shortly to be transferred to Chang, must be left out of this reckoning.

The WAPV already employs Bali men. I believe that in 1904 some 1700 men were recruited and in other years yet more. If Herr van der Loo [manager of WAPV] could realize what evil results this recruiting in large numbers has had for the Bali people he would refrain from estimating what the Bamenda Division can supply. Herr van der Loo has sometimes travelled via Tinto to Bali, but on the basis of the knowledge so obtained he can hardly judge how many recruits are available in the Bamenda Division.

The result of this 'mass-recruiting' is still felt in Bali and is one of the reasons for the dissatisfaction and unrest among the Bali 'vassals'. In 1911, for example, Fonyonga sent over one half of the population of one small village to the WAPV as labourers. The chief is a friend of the Company and received a large 'dash', the sum calculated on the basis of the number of labourers supplied. It was in the interests of the Chiefs to catch as many labourers as possible.

The Bali population as a result of this mass-recruiting is being reduced. The flower of the Bali nation lies on the Cameroon Mountain. In a gang of 300 Balis there were, according to my records, 94 deaths in one year (1912).[3]

According to the last count the Bali and all the 'vassal' tribes only have 4,050 men available. Normally 200 plantation workers could be supplied. The area is already overtaxed by having sent 350 men to work for the Government or the WAPV. In these figures no account is taken of the many free labourers on the estates.

It is clear that for anything over 400, or perhaps over 600 in earlier days, the recruiting of so many more hundreds – up to 1700 labourers – could only have been possible by means of reckless seizures by Fonyonga's so-called soldiers.

Blame for the discontent of the 'vassals' is, of course, conveniently shifted to the Bali ruler and his agents. The 1891 treaty made by Zintgraff, the distribution of breechloaders to Hutter's train-band and other auxiliaries, and earlier extractions of penal labour go unmentioned. Policies had begun to change. As the Governor, Ebermaier, had remarked, in a despatch to the *Reichskolonialamt* in October 1912:

> Bali's importance as a supplier of labour to the plantations on the Cameroon coast declined sharply a long time ago and as a mediator between the Station and other tribes has proved totally unreliable.

The expanded Bali paramountcy called forth by German interests could now be politely dismantled. It had had its uses which were to be recognized by a stipend to its ruler. In providing inducements to chiefs to recruit or impress labour, in the absence of a free labour market, the WAPV was, of course, not alone.[4] While free wage labour had begun to make its appearance on the plantations and in other occupations, labour continued to be supplied by chiefs under administrative compulsion, despite denials of forced labour in the *Reichstag*, for carrying, road-building and maintenance and other public works, and that not only in the German colonies. The railway which, it was long hoped, would at least reach Bagam on the edge of the reduced division, was never extended.⁵ The construction of its first stretch had cost many Grassfields lives.⁶

Notes

1. Reference to these reports is made in Chilver (1967a) 509–510. For the wider administrative background see Chilver (1963) in Robinson and Madden, *Essays in Imperial Government*, 89–139.

2. In addition to O'Neil (1996) in *African Crossroads*, especially 88ff. and the *DKB* reports he cites, see D.A. Vielhauer's reminiscent references in *Fünfzig Jahre Basler Missionsarbeit*, 1936, 55. The attacks on Bamunum (Anong) in 1903–4 are cited as battle-honours for the *Schutztruppe* as are the operations against the 'western vassal towns' of Bali-Nyonga in March and April 1906 (*Stammliste der Offiziere*, 1906, and *Amtsblatt für Kamerun*, 1909). These counted as war-years for military pension purposes.

3. There had been exceptionally heavy mortality in 1912 in the WAPV's Prince Alfred Plantation and a higher than usual rate in the Victoria Plantation: see H. Winkler in H. Stoecker (ed.) Vol. 1, 1960, 274–275. A general overview is provided by Mark DeLancey in his contribution to G. Hartwig and K. D. Patterson (eds.) *Disease in African History*, 1978. See also Rudin (1938), 327–330.

4. See Anne Phillips (1989), *The Enigma of Colonialism*, Chapter 3, for examples from British West Africa.

5. Plans for the extension of the Northern railway beyond Nkongsamba via Chang to Bagam and perhaps beyond to Bamum prompted a survey by Adametz in 1913 (ANY FA 1/936). The first two agricultural settlers had already arrived in Nso', earlier suggested by Glauning in 1906 as a possible area for European settlement.

6. See H. Winkler (1960), for a reference to a medical report by Dr Pfistner of September, 1913, and R.R. Kuczynski (1939), *Togo and Cameroon: a demographic study*, 61.

Figure 15.1: A lasting connection: a visit by Kemner and members of the WAPV management to Fonyonga II in the 1930's.
(From Wilhelm Kemner's photograph album).

Appendix I

The 'Esser Affair'

Esser's book was published by his friend Albert Ahn, in a handsome edition with decorative chapter heads and illustrations, mainly by one Max Raber, early in 1898. It had been preceded by excerpts in the *Kolonialzeitung*, a report in the official *Kolonialblatt* and by lectures. The last included a presentation to the Berlin Geographical Society in 1897, an account of which was published in its Transactions. It was this that got him into trouble.

In December 1898, Esser had been awarded a distinguished decoration, the *Kronenorden* 2nd Class, at the Emperor's own hands – his chief but not his only decoration. This had occurred following Esser's return from another visit of several months to reorganize the WAPV plantations in Cameroon.[1] It had been awarded to Esser specifically for his encouragement of the investment of German capital in the colony. The account of the investiture and of the Emperor's conversation leaked into the press[2] in a version that evidently caused some eyebrows to be raised.

There followed a persistent newspaper campaign against Esser's account of his travels, led by the *Tägliche Rundschau*, and echoed in the *Berliner Blatt*. Esser was accused of giving an untruthful account of the Cunene Expedition and of deceiving the Berlin Geographical Society.[3] The first attacks, one of which describes him as a 'Jewish gentleman', state that Esser could not possibly have covered the ground shown in his route map in the time available to him, as well as making game of some of his anecdotes. Esser and Hoesch, in a brief reply, stated that they had spent 'about two months' in the Mossamedes province and referred their critic to Fournier-Baudach. They did not reply in direct terms to the questions put by their attacker, Dr Hans Wagner, a young political geographer and former pupil of the famous Friedrich Ratzel at Leipzig University – geographer, cultural historian, founding member of the Colonial Society and a prominent National Liberal. Wagner was a contributor to the *Koloniale Zeitschrift*, which he was later to edit for a time. Wagner challenged Esser to state whether he had actually gone beyond the Chella range and had visited Humbe. Accusing Esser of being a liar, Wagner invited him to bring an action for libel. We then learn that Esser challenged Wagner to a duel, of which the police were informed in an anonymous telegram from Königsberg, Wagner's home town. Esser was arrested and condemned to 3 months imprisonment by the courts, a sentence which the Emperor himself remitted to one day's house-arrest. In the matter of duelling the conflict between the Code of the Officer Corps and the law still remained unresolved (Kitchen, 1968; 49–58). This imperial intervention was much discussed and was loyally excused on the grounds that proceedings against Esser had already been started in a military 'Court of Honour'. As Esser held a commission in the Reserve, the *Landwehr* (Cavalry), a court of officers was to determine whether he was a fit person to belong to the Corps of Officers, a matter of honour rather than of military discipline.

The proceedings were so lengthy that Wagner decided to publish a pamphlet restating his charges of scientific fraud and adding further allegations. Among these was the allegation that Esser had curried favour in Court circles by offering the Empress's Chamberlain a large donation to a religious charity in which she was interested, as well as accusations that he had plagiarised the work of others. Vague allegations had appeared in the press about an unspecified 'unpleasant incident' in Cameroon. Rumours were also repeated concerning Esser's private affairs that suggested improprieties in stock-exchange dealings. Wagner represented himself as a lone David, standing up against the Goliath of a capitalist cabal which had penetrated influential Court circles, and which was designed to turn a patriotic colonial movement to its own advantage.[4]

Eventually the 'Court of Honour' was reported to have found against Esser by forty-five votes to five.[5] Esser was said to have admitted that his account of his travels in Angola was 'disconnected' and that he had included information received from others as to the situation east of the Chella range. Had not his book been sub-titled 'Business and Hunting Forays'? According to Wagner, 'everybody' in Mossamedes and Lisbon knew that Esser's unquoted source was the German agent of a Lisbon trading firm. As for Esser's alleged 'discovery' of the true dimensions of Tiger Bay, Wagner claimed that he had merely copied a newly revised Portuguese map, unknown as yet in Germany at the time.

Wagner further claimed that he was approached by an unnamed person who tried to buy him off, saying, 'All African travellers cheat a bit'.

After the verdict of the 'Court of Honour' Esser resigned from the Board of the Colonial Museum, a show-case for colonial interests as well as imported curiosities, and ceased to write for the serious colonial press. However, he continued as Managing Director of the WAPV until 1908, and was in frequent correspondence with the German Foreign Office, active in defence of the company's interests in Bali against the encroachment of the new chartered company, the *Gesellschaft Nordwest Kamerun*. He retained or took up other directorships from his base in the old-established *Schaaffhausen'scher Bankverein*, which remained active in the development of Rhineland heavy industries and mining enterprises.

When Wilhelm Kemner, Esser's successor as managing director of the WAPV, wrote a pamphlet in 1922 celebrating its first twenty-five years and deploring its loss to the British, there is no word of the scandal. Esser still emerges with credit in Kemner's *Kamerun* (1937),[6] despite its author's Nazi sympathies.

None of the attacks against Esser impugned his Cameroon information, apart from his plagiarisms. What, one now wonders, was the hidden prospectus behind 'cheating, a bit,' in his account of the Cunene expedition, which is notably discreet on his discussions with the Mossamedes Company, and silent on earlier exploration and the interests of the South-West Africa Company? Were informants and the longer-term investment plans of important patrons being protected?

Notes

1. See a letter from Governor von Puttkamer to the Colonial Section of the German Foreign Office of 28 October 1898, BAB, R1001, RKA 3499, 78–79. This refers to setbacks in the WAPV enterprise due, he says, to the 'wrong-headed measures' of Zintgraff and the 'unfathomable inactivity' of Spengler. These setbacks had been largely remedied 'by the tireless

activity of Dr. Esser': while profits might not be as high as expected a good financial return was not now in doubt, he believed.

No mention is made in this report of the efforts said to have been made by Esser (*Gouverneursjahre*, 106) to improve the medical care and nutrition of his labourers. Some health items are mentioned in the WAPV report for the financial year 1898–1899, and by 1900 a 'larger hospital' had been built and over 400,000 bananas planted as food (BAB, R1001, RKA 3500, 75). A highly critical retrospective view is to be found in the fortnightly journal, *Koloniale Zeitschrift* (No. 10, 5 May 1900, 137) by then edited by the ever-hostile Dr Hans Wagner. It is from an anonymous contributor who claims 10 years' residence in Cameroon. He blames the 'feverish' propaganda and ignorance of the new plantation companies, in which he detects 'a Semitic element', for circumstances which had led to the need for punitive expeditions. Dr Esser and his first local manager Zintgraff, he avers, did not have the slightest inkling of the tasks of agricultural management. The idea of recruiting Bali, open to question as it might be with regard to what was expected of it, had not been accompanied by the slightest concern for their living conditions, health and feeding by the management: they had 'died like flies', as had the hinterlanders (the Bangwa) brought to Victoria by Conrau. The northern hinterland was now closed. The writer says that he fears that the old style, peaceful, slower but surer plantation development of the past would be ruined if the 'false ideas' and 'bunkum' of the new companies continued to be in fashion. This attack receives some support from a despatch from the interim Governor who described the situation of the plantation workers on the average as 'wretched': unsuitably fed, ill-housed and savagely treated. (Kohler to the *Kolonial-Abteilung*; BAB, R1001, RKA 3227, 51ff. of 1 January 1900).

It would be odd if this anonymous attack came from the long established KLPG, one of whose local managers, Hermann Rackow, had, in an earlier letter to the journal (p. 108), called stridently for higher military expenditure and more effective punitive expeditions to keep the hinterland populations in check and to protect existing investments.

Esser claimed in his WAPV report for 1900 that the earlier high mortality rate among WAPV workers had fallen considerably and that none of the recently recruited Bali had died. In earlier years, he admits, 'we did not have a wholesome food supply for them' and mortality had been 'disproportionately high'. In his report for 1901 he reported a marked improvement over 1900 in the health of indigenous labourers which he put down to the larger area cleared for cultivation, better housing and better feeding arrangements (*Der Tropenpflanzer*, VI, 1902, 304–305).

2. Reported in *Das Kleine Journal* of 18 December 1898, at some length, and more briefly in the *Tägliche Rundschau* of 20 and 22 January 1899.

3. The press campaign against Dr. Esser began with an article in the scientific supplement to the *Tägliche Rundschau* of 26 January 1899 by its

contributor Dr. Hans Wagner. It dealt in particular with the geographical and cartographical points raised by Esser's lecture to the Berlin Geographical Society on 6 February 1897 and later reported in its annual transactions. Possibly it was elicited by the publicity about Esser's decoration. It was echoed in coarser terms two days later by the *Berliner Blatt*. None of the press items we have uncovered so far attack Esser's account of his Cameroon expedition, except in the matter of unattributed passages from a pamphlet by Wohltmann and do not criticise his route map, which contains some misprints, closely follows Zintgraff's and shares its mistaken orientations. These began to be corrected by Moisel in 1903 (*MDS*, 16, 1903, 1–8), on the basis of observations by Besser and Ramsay in the main.

4. Dr. Hans Wagner's pamphlet, dated April 4, 1899, was entitled: *Etwas vom 'Afrikareisenden' Dr. Jur. Esser*. Apart from issuing the results of his detective work to date, which included correspondence with the Mossamedes Company's headquarters and conversations with officers who allegedly formed part of the 'Court of Honour', Wagner had another motive for publishing it when he did. It was to defend himself against the imputation that he had funked the duel with Esser.

A later piece by him in *Tägliche Rundschau*, of 6 October 1899, returns yet again, and with virulence, to the dangerous influence of capitalist 'agitators' in high society and court circles. Can we account for Wagner's continued pursuit of Max Esser (which had brought him to wider notice) in political terms? His pamphlet comes shortly after the agitation that followed the notorious Stock Exchange coup of the founders of the *Gesellschaft Süd-Kamerun,* in particular their feats on the Brussels Stock Exchange at which participation rights were made available. Huge profits were allegedly made by Scharlach and Douglas and by participating Belgian and German banks. This aroused widespread criticism especially among members of the Pan-German League. Wagner's attitude to concessionary companies was set out in an article in *Der Weltmarkt* (No.15, 1 August 1899). In this article exceptions are made in favour of mining concessions in otherwise useless lands, borderland concessions to prevent encroachment by foreign chartered companies, and some railway investment. He again mentions Esser 'the stockbroker', together with Max Schöller, the GNK's front man, as persons who hide their operations behind princely chairmen.

Wagner himself was not to escape attacks from the more xenophobic Pan-Germans who went so far as to accuse him of connections with the '*System Scharlach*', as the Africanist Professor and explorer Siegfried Passarge called it, and to impugn his independence as a journalist after he had begun to edit the *Koloniale Zeitschrift,* as its issue of 24 April 1900, relates. Esser was not, however, Wagner's only target. The Governor of German East Africa, von Liebert, was next pilloried together with his supporters in the press, in a pamphlet which made Wagner more enemies: see Wagner (1900). For a brief account of how

Scharlach was viewed by his opponents see Woodruff D. Smith (1978), 160–161.

5. The decision of the 'Court of Honour' was reported in early August, 1899, issues of the *Tägliche Rundschau* and other newspapers. Its proceedings do not survive. The greater part of the Prussian Army's military court archives was lost in the Second World War (personal communication, Ms. Zandeck, Bundesarchiv-Militärchiv).

6. See Wilhelm Kemner (1922) and Kemner (1937), *Kamerun*, 135–137.

APPENDIX II

The 'Fetishes' and the Esser Collection at the Linden Museum

Ute Röschenthaler

This appendix deals with what we know of the acquisition history of Esser's ethnographic collection and his 'Bakundu fetishes'.

In his Chapter 6, on the expedition to Bali, Esser tells us that he had collected four 'fetishes' from 'the Bakundu', evidently a rather loose ethnic description. One of these came from Ikiliwindi, a Bafo (Bafaw) village, and was exchanged for an accordion. The three others were acquired further north-east in Bulo Nguti, another Bafo village, after lengthy negotiations with the chief and 'the sorcerer of the village'. It is not shown on modern maps under this name, but that of Kurume[1] Esser does not describe the exact circumstances of their acquisition; we do not know where he first saw them, and, still less, how they were used.

Esser realised, however, that these sculptures were remarkable specimens of their kind, and that the Berlin Ethnographic Museum would not yet be in possession of similar ones. His remarks lead one to expect that they would be found at the Berlin Museum; however, they are not. The Berlin Museum has parts of Zintgraff's collection, including pipes, spears, and pottery from Bali (Luschan, 1897); some other items collected by Zintgraff are in the Linden Museum. It is here that we find Esser's collection. The acquisition lists of the Linden Museum show that they had received from Max Esser 210 ethnographic items, among them the Bali royal wives' spears and *nggwasi* described in his book. The Bulo Nguti 'fetishes" are there too, though mistakenly identified as 'Bali'. How did they get there?

Acquisition history

The Linden Museum opened in October, 1897, to the applause of the great and the good.

It was founded by Graf von Linden under the auspices of the *Verein für Handelsgeographie* at Stuttgart, a scientific society of some 900 members which included the King of Württemberg. It needed more ethnographic collections, so, on 30 November, 1897, Graf von Linden addressed a letter to 'Max Esser, Kamerun, Deutschwestafrika', inviting him to cooperate with the Museum by donating a collection of basic utilitarian and ornamental items collected during his Cameroon travels. Weapons, he writes, and hunting gear, textiles, drums, tools, pipes, dance masks, and fetishes, calabashes, ornaments as well as photographs of places and people would be very welcome since 'the simplest everyday objects will give a full picture of the cultural development of a people'. Esser was already personally known to Graf von Linden, since he had given a lecture to the *Verein* in Stuttgart in the winter of 1897.

A year passed before Esser replied, on 20th December, 1898, explaining that he had just returned from a visit to Cameroon. Yes, he would be glad to supply the Museum with some large wooden fetishes and *ethnographica* including pipes and weapons. With his letter he enclosed a copy of his book, fresh from the press.

On 3 January 1899, Esser forwarded a considerable part of his collection to Stuttgart accompanied by a letter addressed to the King of Württemberg through his council of ministers. A copy also went to Graf von Linden, who, as etiquette required, was supposed to take the collection over from the King. A translation follows:

> For the attention of His Excellency, *Legationsrat* Freiherr von Gemmingen.
> I have the honour to refer to the letter of December 23rd, 1898, from the royal council in which you kindly conveyed His Majesty's gracious approval to my placing, at the disposal of the local ethnographic museum, the fetishes illustrated on p. 112 of my book and in addition to these fetishes, the ethnographic items listed below:
>
> – Calabash drum (Banjang)
> – Two palaver drums (Kamerun [Duala])
> – One big King's drum
> – A dance bell
> – A drinking cup
> – A blue Hausa garment (a present from King Garega)
> – Three dance rattles
> – One helmet
> – A pair of sandals
> – Four musical instruments

- Fourteen black Bali caps (of woven grass)
- Four coloured Bafut caps
- Two carved wooden dishes
- One iron bell (iron processed in Bafut)
- Ten Bali knives
- One Herero sword
- One Bali sword
- One Bafut sword
- One whip (of elephant hide)
- Three wood projectiles (for hunting antelopes)
- One Bali basket, round
- Two Bali baskets, rectangular
- Six bags made of woven grass
- Three ammunition bags (of fur)
- Seventy-five Bali personal ornaments, various (of metal and wood)
- Two palm wine calabashes, with burnt-on designs
- Ten antelope and gnu [sic] horns
- A packet of Bali women's garments (38 pieces)
- One dance cap
- One wild-cat skin
- One Kamerun [Duala] staff
- A collection of fifty pipes with different bowls

Might I draw your attention in particular to the collection of pipes, since no other museum on earth possesses such an extensive collection? The Bali, being passionately addicted to smoking, not only smoke their pipes while drinking their palm wine morning and evening, but, being so attached to them, keep them alight during battle. The Bali have reached a noteworthy level of skill in pottery and have invented forms which are remarkable in their fantasy and variety. These smoking pipes, every single one of which is different, mostly represent human heads, and repeat the human form in a variety of postures, gestures and stylizations. I draw attention, too, to their peculiar ornamentation of monkey heads, lizards and snakes and to the ingenuity of the artists who can portray such an array of nature in such a small space. The grotesque impression of the heads is enhanced by skilfully decorated pipe-stems overlaid with tin foil.

Also of interest are the Bali crafts of basketry and weaving as evidenced in the baskets and bags sent. The craft of loom weaving is well developed, as can be seen in the nicely patterned travelling-bags of raffia cloth. I would like to draw special attention to the comparatively high level of smith-craft among the Bali and neighbouring peoples, which is finely exemplified in the weapons and tools despatched. The rest of the collection displays the usual local inventory of the Blacks, objects for use as well as adornment. All these are not lacking in some peculiar feature, some highly interesting, and without them the ethnographic picture of the tribe would be quite incomplete.

In conclusion may I express my gratification at the great honour His Majesty has bestowed upon me by his gracious words of recognition of my travels and gracious reception of the attached collection ...

Signed: Dr Esser, African explorer and Lieutenant in the *Landwehr* (Cavalry).

The collection arrived on 5 January 1899 in Stuttgart. Its arrival and scientific importance was reported to the King by the responsible diplomatic officer and court chamberlain, Freiherr von Gemmingen; the King was delighted and minded to confer a decoration on Esser.

Unfortunately, ten of the finest and largest earthenware pipes were broken in transit. Moreover only 41 had arrived – of the 50 listed. In his letter of thanks (8 January 1899) for the 'brilliant collection', Graf von Linden allows himself to ask Max Esser to replace the broken pipes and add those missing. In a light-hearted and friendly letter he comments further on the collection:

> 'Do we understand from the euphemism 'Bali women's garments' the *guassi* [sic] mentioned in your book? . . . they look rather like South Sea women's garb, also to be seen in East Africa . . . Now for something else. Perhaps it was rash of you to send me your well-written book, although I assume you did so on purpose, and I have interpreted it in terms of your good nature. I have read it carefully and of course it aroused my greed for more things. But I shall restrain myself and only mention an unintended *gaffe* of yours. If one is giving a ruler presents from Bali it is surely *de rigueur* to include some of the most beautiful spears of the wives of His Black Majesty Garega? Nothing else, but do be kind and – please – send us some of these Bali iron products to round off the collection appropriately. According to a confidential report you should be receiving such a beautiful souvenir of Württemberg that it should not be too hard for you to add to it.
>
> In that case, I shall, of course, announce it officially. The dear ruler of your house will not, I'm sure, enjoy queening it among all these witnesses to African fantasy, so, you see, I am ready to help you out of this quandary. One more word about those humdingers of fetishes. Please have a look to see whether you are still in possession of the right hand of the male fetish – it is a fresh cut. And what is the meaning of his testicle being split? Is it just an intended flight of fancy, or is it practised? The fetishes whispered in my ear that they had got so fond of their comrade from Ikiliwindi that they can't bear to be without him. Could you not release them from their anguish by re-uniting them? . . . Do you have any collections from Togo? His Highness is interested in collections. I had a long audience with him yesterday . . .

On receipt of this letter Max Esser decided to send the rest of his extant collection to the Linden Museum and sent a telegram to this effect. On 11 January, he wrote to Graf von Linden as follows:

> Dear Graf, highly esteemed friend and patron, On receipt of your letter of the 8th of this month I decided to supply the museum with the rest of my collection and accordingly I yesterday despatched a packing case and package containing the following objects:
>
> – Twenty-four pipes
> – Six Bali women's spears with broad iron blades, some chased
> – Eight fighting spears
> – One rattle-spear, for dancing

- One scout's spear
- One hexagonal basket
- Six bracelets, bronze
- One carved horn
- One hair pin
- One ivory signal flute
- One dance bell
- One stately calabash of King Garega's which he gave me on my arrival in Bali (a stopper is enclosed with it)[2]
- One palm wine calabash covered in wicker work
- One carved sceptre, as carried by Garega's crazy scouts
- One Bali shield of elephant hide, with designs

I sincerely regret that a number of pipes were broken. This is all the more to be regretted since these pipes, because of their fragility, are difficult to export to Europe. I had intended to acquire a sample collection and successfully managed to transport no less than 300, all of different types, from Bali to the coast. Except for about 60 all got broken. Nevertheless no other museum in the world will have such a well-assorted collection of pipes from a single tribe.

On the question of 'garments' you are right, they are the *guassi*. Those with the well-interlaced and coloured crests with points are worn over the buttocks and the simpler ones without points are worn over the pubic region, both of them hanging from a cord round the waist. These *guassi* are only worn for festivities and woven anew for each event from sweet-smelling coloured grasses so that these items of apparel look pretty and smell fragrant.

The fetish from Ikiliwindi was stolen on the way to Europe and despite all endeavours has not been rediscovered to this day.

It has been a pleasure to learn from you that His Majesty in the audience of the other day has confirmed an interest in collections as well as the indication that with your kind support I will soon receive a beautiful souvenir of Württemberg.[3]

Most dear Graf, please accept my sincerest thanks for this. Please do me the favour of remembering me to your wife. I am inclined to agree with you that my future wife will be grateful that you have removed the collection from me, since the fetishes and pipes would not be the most agreeable co-denizens of a home.

With my greatest respect and gratitude.

Dr Esser

PS You asked me about the split testicle of the male fetish. I must admit that this did not strike me at the time. Garega has some eunuchs who live in his palace but I never got any information as to whether this was due to a punishment or whether they served as guardians of his wives. Garega was very suspicious and reserved about everything concerning his women, for no apparent reason.'

This time the pipes arrived unbroken. In his letter of thanks (15 January 1899) Graf von Linden says that the collection would be displayed as soon as possible, and asked a few more questions about the collection. Esser replied briefly on 20 January 1899, in the following terms:

> ... May I express my warmest thanks for your letter of the 15th and my gladness that the collection arrived intact? I have not found the hand of the fetish. The

small calabash served as an ammunition [*sic*, as a gunpowder?] container in our time. Originally, the Bali always carried the cubeb pepper, which they combined with Kola nuts when drinking palm wine, in such containers. A successful elephant hunter is allowed to wear its cut-off tail around his waist for a month. The hexagonal basket was supplied with a wooden base which has probably been lost too. Unfortunately I am not in possession of any photographs of men and women. Garega is shown in my book. I shall gladly recommend the Museum in research circles.

The Linden Museum acquisition books of 1899 list the Esser collection more or less as it is set out in his letters. Many of the items listed are no longer to be found there, partly as a consequence of the Second World War, and partly as a result of exchanges for other objects.[4] Only two of the *nggwasi* and two of the women's spears can now be found there. The women's spears are impressively tall, 2.08 and 2.04 m respectively by comparison with the Bali men's spear, of 1.88 m in height.

Grassfield pipes are well documented and aroused the interest of early museum ethnographers. Luschan, for example, admired the skill shown in their production and refers to the large collection of Bali pipes at the Berlin Ethnographic Museum in a book which documents the first colonial exhibition at Berlin-Treptow in 1896 (Luschan, 1897). The many small objects collected and labelled as 'adornments' bear witness to Esser's lasting interest in fashion in all its manifestations, as well as his love for evidence of hunting. The collection, for example, includes a leopard's tooth, a number of horns, small sticks (lip plugs?), ivory and grass bangles, an elegant hairpin, bracelets with metal elements, fruit shells, and wooden staves. Many of them are clearly not ornamental. Some of the horns, for example, contain a mixture of undefined substances. Almost all of them were collected *at* Bali but do not necessarily derive from Bali itself, given the considerable exchanges of artefacts and prestige goods in both regional and longer distance trade. Garega, for example, presented Esser with a 'Hausa' garment – perhaps Esser's gloss – the cloth for which might well have come from a variety of sources to the north-west or from European factories on the Benue by that date. Hausa traders were, according to Zintgraff, von Stetten and Hutter, already well within easy reach. (See also Wilhelm 1981).

Other objects were collected *en route* – the 'Bakundu fetishes' and a Banyang drum.

APPENDIX II 177

Figure AII.1: Royal Wives in Bali-Nyonga, 1907–9, one carrying a spear. (Ankermann Collection No. 310. Reproduced by courtesy of the Museum für Völkerkunde, Berlin)

Figure AII:2: Two of the famous *nggwasi,* 'Women's garments' (10942, 10943). (By courtesy of the Linden Museum)

178 APPENDIX II

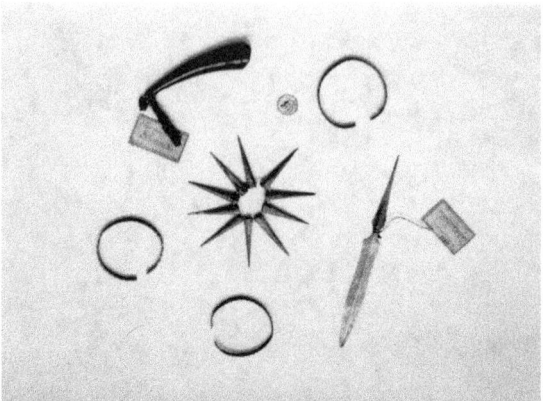

Figure AII.3: A 'hairpin' (hair ornament), 'fetish' (medicine) horn, three metal bangles, and a necklace made of brass cartridge cases (10935, 3682, 3263, 3658, 3662, 3698). Items from the Esser Collection at the Linden Museum. (By courtesy of the Linden Museum)

Figure AII.4: The three 'fetishes' from Bulo Nguti presented by Esser to the Linden Museum (3395, 3394, 3393). (By courtesy of the Linden

The Fetishes

The 'Bulo Nguti' sculptures are exceptional. Most known sculptures from the area are small by comparison; the height of the female figure is 1.35 m and of the male one 1.38 m 20 cm of which is represented by the little stool it stands on. The large, and quite heavy wooden board is 1.71 m high by 32 cm wide and might have been even higher, since its lower edge has been eroded by termites or decay. The missing figure from Ikiliwindi was definitely the largest, about 2m high. The small round stools shown in Esser's plate are quite reminiscent of simpler Grassfields styles.

Keith Nicklin (1996, 348) and Peter Valentin (1972) confirm that not much is known about Bakundu sculpture. In a short survey of references in the literature on African art Valentin shows that the Bakundu sculptures usually described as 'fetishes' are much smaller: indeed the largest in the Basel Mission Museum is a mere 25 cm high. These seem to be those described by Nicklin, and also Northern (1984, 187), as owned by one of the Bakundu secret societies and used in the administration of judicial oaths, while the larger ones were 'used to enforce dispute settlements', and carried on the back in dance or procession on ceremonial occasions. Even less is known about Bafo and Balong sculptures.

There are a few other huge 'Bakundu fetishes' in ethnographic museums. Bernard Gardi (1994) indicates that the Basel Mission Museum had owned another large sculpture, not mentioned by Valentin, as it was considered an unauthentic intruder into the Bakundu series. It was sold to the Barbier Müller Collection in Geneva. This 2 m high sculpture came to Basel in 1898, and had supposedly served as a pillar in the secret society meeting house of the Balong village of Ndo, north-east of the mission station of Bombe, founded by Missionary Lauffer in 1896/7. Lauffer acquired it in exchange for 'a pile of goods' and for sending the Balong village a teacher. This sculpture, referred to as 'fetish Dikoki,' reminds one of the stolen Ikiliwindi sculpture, which looks older and is slimmer.

There is another huge Bakundu sculpture in the Berlin Ethnographic Museum (III C 10026) collected by the German colonial officer Conradt in Bonge village in 1899 (illustrated in Krieger, 1978, pl. 145; Sydow ed. Kutscher 1954, pl. 49D; Beumers and Koloss, 1992, 308; ed. Phillips 1996, 348). It is a male figure 1.72 m in height. Its entire right side is painted brown, the left side white. Its head is encased in a diamond-shaped frame resembling that of Esser's board carving and also found in other Bafo sculptures or headdresses (Krieger, 1978, pl. 155).[5] Lieutenant Lessner makes no mention of such 'fetishes' in his account of the Balue but reproduces a photograph of 'Ngolo idols' (1904: 338). It shows a large board, carved in relief, evidently carried outside by the owner to be photographed in the front of a house. It looks like Esser's in size and is divided into sections. There are no carved heads. The central section is a carved circle framed in a square. Above and below it are various small ornamentations and an anthropomorphic figure in each section. Beside the board stand two anthropomorphic carved figures of about half the size of the board. They are like those collected by Esser but their arms are differently disposed.

Esser does not describe the context or use of his 'fetishes'. In the 1880s mention is quite often made of large sculptures, again without describing their uses. The Polish explorer Rogozinski (1884, 134) had seen a sculpture in the house of the 'King' of Mukonye village. It stood in front of a pillar in a ready-to-spring position, painted in black and white. The Swedish trader and planter Valdau (1886, 46) describes a clothed sculpture in the village of Bakundu ba Bakäa. He writes:

> In some of the houses there were human figures carved out of wood on the top of poles ... in others free-standing human images, about 1.5 m in height, which are dressed in a costume similar to that worn by the *Ekale* societies.

During his first visit to the Bakundu the German Government's representative Max Buchner mentions man-sized posts surmounted by carved heads in Bakundu ba Nambele (1914, 181). On his second trip to the Bakundu in 1885, Buchner describes the interior of a chief's house where he lodged, more evocatively. This was at Malende, a Balong village on the route between Mt Cameroon and Bakundu ba Nambele, where the Black American missionary Richardson had stayed with his wife.

> [Surrounding the two poles to which Buchner's hammock was hitched] ... were grouped a whole museum of ethnographic-religious oddments. In front of each pole stood a malicious looking idol of human height carved in wood. His roughly carved limbs were hung with devotional objects such as beaded strings, small roots and beans on threads, dried bird's heads, moth-eaten plumage still a-glitter, rustling branches of sacred foliage, power-filled sprays of fruiting pods, little European mirrors in African frames which magically reflected the light from the door in the dark room, small pumpkins filled with electuaries attractive to good spirits and much more of the same. Curious netting and multiple knotted strings blackened by smoke wove their way through the whole space. Among them were arranged Gods of lesser calibre, male and female, with grotesque shapes, oddly arranged hair growths and little snails inserted for eyes.

So feeling well protected, Buchner prepared for sleep, pondering how he might transfer these magical objects to museums at home, to the joy of German professors.[6] Heinrich Balz, in describing the Bakundu sculpture in the Berlin Ethnographic Museum (Beumers and Koloss 1992, 308) assumes that such fetishes were 'originally not naked but clothed.' The hole in his hand suggests he held something – a spear or an axe. The 'fetish Dikoki', illustrated by Lauffer (1898) and Gardi (1984), wears a dark-coloured long skirt, and holds a carved society staff in his hand which signifies his membership of it. Zintgraff, too, saw and coveted some man-high sculptures in the Balong village of Baduma (1895, 83).

None of these accounts, unfortunately, describe a carver at work. Nor do we know which figures were made by freemen, which by slaves – among them the Bayong, Mbudikum or Mbrikum, to give them the terms by which the slaves from the distant interior were known to the traders of Duala and Calabar (Warnier, 1985, 173–177).

Other sculptures were publicly burnt. Conradt was at loggerheads with 'fetish priests' and paints an awesome picture of them in his account of 1900. His Bakundu live to the west of Kumba, but the term Bakundu was used by early travellers to cover the region north and south of Kumba which included several ethnic groups, Bakundu, Balong and Bafo among them. He does not mention 'Bonge' where he collected his big Berlin 'fetish': it might well refer to the Mbonge ethnic group. Conradt suspected that the 'fetish priests' were inciting the people against the Government Station, against Europeans in general, and were making laws that nobody should work at the Station and were confiscating the wages of Station workers. He then imprisoned three of the most influential 'fetish priests' of Kumba, and gave orders that all 'fetish objects' were to be confiscated and destroyed. Thus, he says, the power of the 'fetish priests was almost completely broken and the entire population could heave a sigh of relief' (1900, 36).

According to early reports, as we saw, large sculptures were found among various ethnic groups, including the Bafo, Balong, Bakundu and Ngolo. The reports suggest that the sculptures were part of a shrine in the community hall or 'fetish house' at the centre of a village and that some villages had up to three. The houses contained the symbols of the secret societies. In all probability, the sculptures belonged to one or other of the numerous secret societies or cult agencies of the region. These were commonly disseminated from one village to the next, regardless of ethnic boundaries. In the course of this dissemination their names were often changed –

usually the sculptures bore the same name as the societies to which they belonged – and the appearance of the figures associated with them was modified a little to meet a village's preferences. Given the dispersal of similar societies the presence of large sculptures among the Bafo, Balong, Bakundu and Ngolo can be readily explained (Röschenthaler in *Tribus*, 1999). In a manner similar to the practice among the Ejagham, the big sculptures can have represented a cult agency set up to protect the community against all evilly-disposed transgressions – theft, witchcraft and sorcery. The Ejagham still own such cult agencies (*ajom*). Their shrine usually stands in the centre of the village. At the turn of the century they also had included anthropomorphic sculptures. Additionally each member owned a set of smaller sculptures that helped their owner to solve the problems of their clients (Röschenthaler 1993).

Esser's 'fetishes' must have been among the last still existing in 'Bakundu'. 'The Bakundu' do not seem to have continued to own large sculptures for long after the turn of the century. When I presented some photocopies of Esser's sculptures to some of the chiefs in Bafo, Bakundu and Balong villages in 1999, they were surprised to learn that such large sculptures had been found in their villages. None of them had ever seen one. Some spontaneously suggested that they must have come from the Grassfields, rather than from their own area. Except for the clan shrines, which represented the three Bafo clans, there were no longer any shrines to be seen in the villages. However, the various secret societies continued to exist. New sculptured objects had often been bought or commissioned by the Bakundu from the Bafo and Balong, the latter of whom were, towards the end of the century, disposed to sell their products to traders and travellers, and even made expeditions to Duala to dispose of them. Bohner (1896, 382) describes the Balong as 'one of the few tribes who own carved idols or figures of fired clay'. On their first visit to the Balong, the Basel missionaries were allowed to see and touch the figures, but they were unable to acquire a single one of them. In the course of its activities, the Mission started denouncing the practices of the secret societies and soon got into conflict with them. One day, a faithful Christian even managed to carry away a 'fetish' to the missionaries. To the surprise of the public, he did not die immediately afterwards as a consequence of the strong powers it was assumed to have. After this incident in Muyuka, Bohner writes, the 'worship of idols' was shaken.[7] At first, a few men started to sell their idols secretly to the Mission teachers, but before long they began to do so publicly. Any of them who had the occasion to steer for Duala sought to dispose of such figures by sale, and these wandered to Europe in packing cases.

Notes

1. According to Nfon E.M.A. Epie, Chief of Kurume, Bulo Mbu Nguti was a title of a chief of Kurume. The British Assessing Officer, R.W.M. Dundas, confirms that Nguti is a title-name of the Difon secret society, which only heads of the founding families of the Bapewan clan of the Bafo are entitled to use: see his *Assessment Report for the Bafaw*, 1922 (Buea National Archives). The designation Kurumen appears, in place of Bulo Nguti, on a map accompanying a report by Governor von Puttkamer (1901).
2. This calabash is on display at the Linden Museum Stuttgart.
3. It rather looks as if Esser did not receive the promised decoration in the end; his name does not appear on the lists of recipients of Württemberg decorations of this period (personal communication, from Herr Braunn, *Hauptstaatsarchiv Stuttgart*). This might be attributed to the 'Esser Affair' becoming known to the King of Württemberg. The Museum correspondence includes newspaper cuttings relating to it.
4. In 1971, for example, the University of Mainz had exchanged a collection of Hindu Kush objects from its teaching and research collection for a variety of objects from the Linden Museum. These included seven items from the Esser Collection: a musical instrument (sansa), a skin armlet, a horn, a necklace, a dance rattle, and two woven caps.
5. Missionary Lauffer *(Heidenbote* 1898) suggests that Dikoki is wearing a top hat, the shape of which indicates to the author that the carver must have had contact with early Europeans or those who knew them. The sculpture in the Berlin Ethnographic Museum reminds Keith Nicklin of another type of European hat: a sea captain's or officer's hat. Whether these types of headgear were directly inspired by European models or not, they usually indicate the rank of its wearer, and transfer something of the original owner's power to its new possessor. A hierarchy of clothing was also quite common in secret societies. Only the chiefs were allowed to wear hats.
6. Buchner did not dispose of any Bakundu fetishes to the Munich Ethnographic Museum, of which he became a director after his time in Cameroon (personal communication, Maria Kecskesi).
7. Accounts of 'the burning of the idols' in the first wave of enthusiasm for the Basel Mission around Bombe, and the trade in them is referred by Schlatter (1916, vol. 3) to contemporary reports in the Basel Mission's journals, *Der Evangelische Heidenbote*, (1896, pp. 6 and 48), and *Evangelisches Missions-Magazin*, (1898, pp. 372–385).

Map 1

This sketch map has been redrawn from the fold-out route map appended to Esser's book on a base supplied by the modern maps issued by the Centre Géographique National and greatly reduced in scale. It shows only those sites which can be more or less securely identified.

As Esser relates many villages had moved since Zintgraff's last passage through them and some have certainly shifted again around the approximate sites shown or merged with neighbours and disappeared or become known by other names. Other groups, Mbonge, Basossi and Bakossi for example, have since moved closer to Esser's route which seems to have lain a little to the east of the first stretch of the modern road from Kumba to Mamfe.

Bulo Nguti is not shown on modern maps. But Chief Epie of Kurume (Bafo) could recall that an earlier chief was called Bulo Mbu Nguti: see Appendix II. Likewise Esser's Miyimbi recalled a predecessor of the chiefly line of the Banyang village of Tali. According to Professor Mbuagbaw, who has recorded its history, Tali was reportedly a name given it in German times.

We know that Banti (not shown) between Sabe's and Bamessong (modern Ashong) must have moved, perhaps to its present site, since the extensive ruins of old Banti were discovered by British officers mapping the divisional boundary between the former Bamenda and Mamfe Divisions in the 1930s.

Esser, as we know, disclaimed any special expertise as a topographer and states that he and Hoesch relied on Zintgraff for such matters during their Cameroon expedition. Zintgraff's earlier route is shown on Esser's route map with a few of his deviations from the 1896 route. In the acknowledgements in his foreword to those who helped with his maps Esser mentions no names, and the map itself only bears the name of a Berlin lithographer.

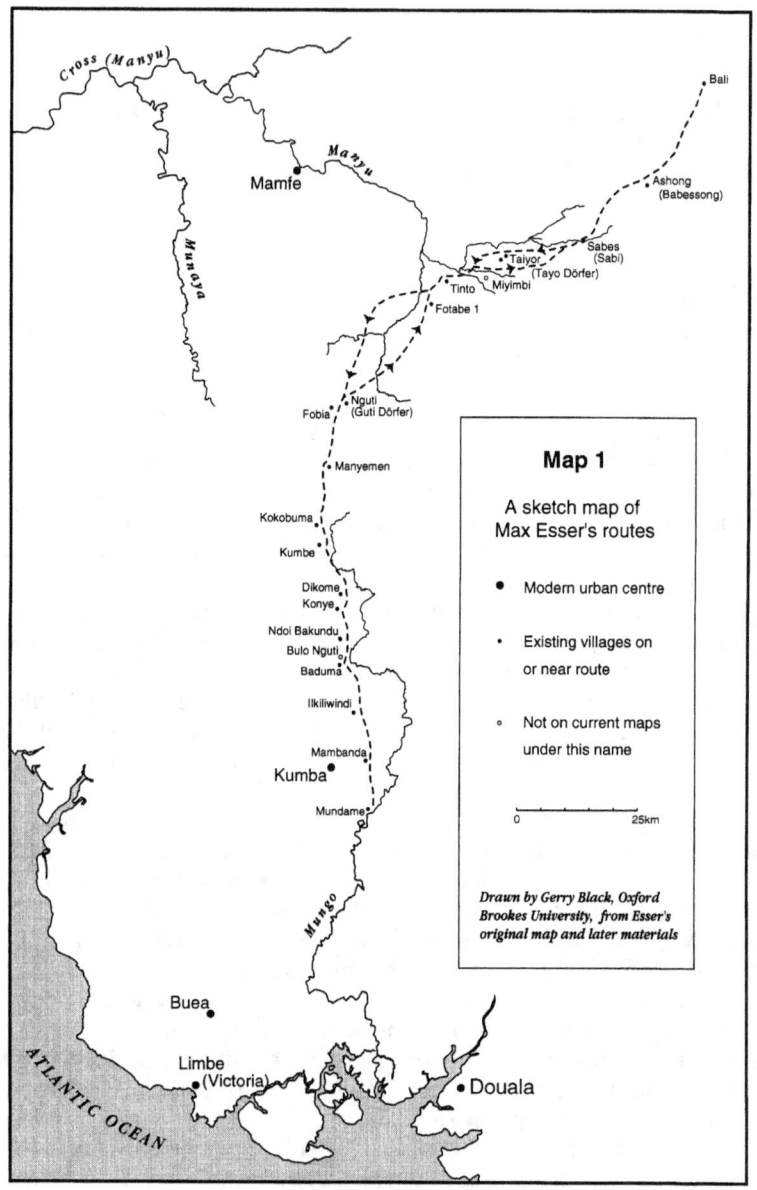

Map 2

This is not a linguistic map, properly speaking. The typography used does not indicate the interrelationships between the languages spoken along Esser's route or mentioned in the text. Had we attempted to do so we would, for example, have shown the Littoral Province congeners of the little islands of Barombi speakers. We have retained the linguists' Oroko label for a large central group, which includes the Bakundu, north and south, and the Mbonge since it has become a common designation. Uncertainties in classification remain and old assumptions have recently been challenged. So we have merely indicated the more important mother-tongues, and shown distinct local languages and dialects in smaller type. The south-west corner of the North West Province is a mosaic of Grassfield languages belonging chiefly to the so-called Momo (e.g. Mogamo) and Nggemba (e.g. Bafut and Mankon) groups. The common language of Bali-Nyonga is derived from another Grassfield language found to the east; Bali-Nyonga also contained speakers of a quite different language from Adamawa which had rather more speakers in other Bali settlements. Vocabularies of distant languages, Wute and Tikari for example, as well as some from the more easterly Grassfields, could be collected as late as 1960 from elderly persons. These confirm its original composite nature as well as its readiness to incorporate strangers.

The language limits shown must be taken as approximate. Modern administrative centres have been shown to orient the reader. We have gratefully used as our base and adapted the administrative maps of national languages by province and division compiled by Breton and Fohtung (1991) and published in Paris and Yaounde, as part of a cooperative project promoted by the French Cultural and Technical Cooperation agency (ACCT).

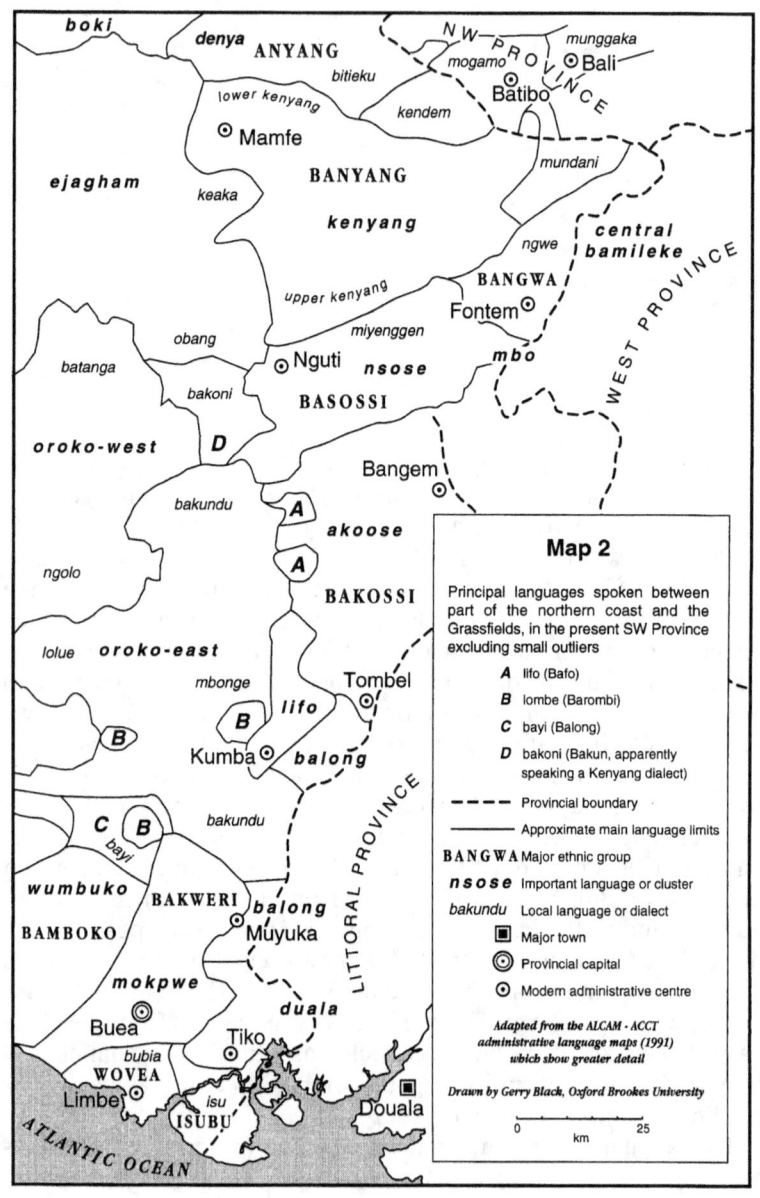

Bibliography

Africa Bibliography Series: West Africa, 1958– (subsequently enlarged and published annually for the International African Institute, now by the Edinburgh University Press).
Anon., 1928. *A. Schaaffhausen'scher Bankverein, 1848–1928* (pamphlet). Köln.
Anon., 1917. *Briefadel Taschenbuch*. Gotha.
Anon., 1920. 'Der Familienname Esser', *Familiengeschichtliche Blätter,* p.45.
Ardener, E.W., 1956. *Coastal Bantu of the Cameroons* (Ethnographic Survey, West Africa. 11). London.
—— 1962. *Divorce and Fertility: an African Study*. Oxford.
—— 1996. *Kingdom on Mount Cameroon. Studies in the History of the Cameroon Coast, 1500–1970.* Providence & Oxford.
—— (forthcoming). *Oral tradition and administrative identities.*
Ardener E. W., S. G. Ardener & W.A. Warmington. 1960. *Plantation and Village in the Cameroons*. Oxford.
Ardener, S. G., 1968. *Eyewitnesses to the Annexation of Cameroon 1883–1887*. Buea. (Reprint by Friends of the Buea Archives, Oxford, 1996).
Austen, Ralph A., 1977a. Slavery among coastal middlemen: the Duala of Cameroon, in S.Miers & I. Kopytoff (eds.), *Slavery in Africa.* Madison WI, 305–333.
—— 1977b. Duala vs. Germans in Cameroon: economic dimensions of a political conflict. *Revue française d'histoire d'outre-mer* 64 (4), 477–497.
—— 1983. The metamorphoses of middlemen: The Duala, Europeans, and the Cameroon Hinterland, c.1800–c.1960. *International Journal of African Historical Studies* 16 (1), 1–24.
—— 1986. Cameroon and Cameroonians in Wilhelmian *Innenpolitik*, in K. Ndumbe III (ed.), *Africa and Germany from Colonisation to Cooperation, 1884–1986.* Yaounde, 204–226.
—— 1996. Mythic Transformation and Historical Continuity. The Duala of Cameroon and German Colonialism, in Fowler, I. & D. Zeitlyn (eds.), *African Crossroads.* Providence & Oxford. 63–80.

Austen, Ralph A., & Jonathan Derrick. 1999. *Middlemen of the Cameroons Rivers.* Cambridge.
Ballhaus, J., 1968. Die Landkonzessionsgesellschaften, in Stoecker (ed.), Vol. 2, *Kamerun unter deutscher Kolonialherrschaft.* Berlin. 130–162.
Balz, Heinrich, 1984. *Where the faith has to live. Studies in Bakossi Society and Religion.* Basel & Stuttgart.
Barreteau, D., E. Ngantchui & T. Scruggs (eds.), 1993.*Bibiographie des langues camerounaises.* Paris.
Barth, Boris, 1995. *Die deutsche Hochfinanz und die Imperialismen: Banken und Außenpolitik vor 1914.* Stuttgart.
Bastian, Adolf, 1884. *Der Fetisch an der Küste Guinea's.* Berlin.
Baumann, H. &. L.Vajda, 1959. Bernard Ankermanns völkerkundliche Aufzeichnungen im Grasland von Kamerun 1907–1909. *Baessler-Archiv* N.F. 7, 217–307.
Berghahn, V. R., 1994. *Imperial Germany, 1871–1914.* Providence & Oxford.
Besser, Lt. von, 1898. Bericht über die Expedition zur Festsetzung der deutsch-englischen Grenze. *Mitt. aus den deutschen Schutzgebieten* 11, 177–183.
Beumers, Erna & H-J. Koloss, 1992 . *Kings of Africa.* Maastricht.
Birmingham, David, 1999. *Portugal and Africa.* Basingstoke & New York.
Böckner, G., 1892, 1893. Streifzüge in Kamerun. *DKZ,* N.F. V (10), 137–139; VI (1, 3, 6), 7–9, 35–37, 74–75.
Bohner, Missionar, 1898. Skizzen aus der Kamerun Mission. *Evangelisches Missions-Magazin,* 372–385.
Bouba, Aissatou, 1996. 'Lauter breite Negergesichter'. Die Darstellung der äußeren Erscheinung einiger nicht-moslemischer Ethnien aus Deutsch-Nordkamerun in der Vorkolonial- und Kolonialzeit. *Paideuma* 42, 63–83.
Breton, Roland & Bikia Fohtung, 1991. *Atlas administratif des langues nationales du Cameroun.* Yaounde & Paris.
Brose, M., 1897. *Die deutsche Kolonialliteratur von 1884–1895.* Berlin.
Buchner, Max, 1887. *Kamerun: Skizzen und Betrachtungen.* Leipzig & Berlin.
—— 1914. *Aurora colonialis. Bruchstücke eines Tagebuchs aus dem ersten Beginn unserer Kolonialpolitik 1884/85.* Munich.
Bufe, Missionar, 1913. Die Bakundu. Volkskundliches Material über ihre Sitten und Rechte. *Archiv für Anthropologie* (Braunschweig) 12, 228–239.
Burton, R. F., 1862. *Abeokuta and the Cameroons Mountain: an Exploration.* London.
Cecil, Lamar, 1989. *Wilhelm II: Prince and Emperor, 1859–1900.* Vol. I. Chapel Hill & London.
—— 1996. *Wilhelm II: Emperor and Exile, 1900–1941.*Vol. II. Chapel Hill & London.
Champaud, J., 1966. L'économie cacaoyère du Cameroun. *Cahiers ORSTOM,* Série Sciences Humaines III (3), 105–125.
—— 1983. *Villes et campagnes du Cameroun de l'ouest,* maps. Paris.
Chem-Langhëë, B. (ed.), 1995. Slavery and Slave-Dealing in Cameroon in

the Nineteenth and Early Twentieth Centuries. *Paideuma* 41, part 2, 95–272.
Chem-Langhëë, B. & E. S. D. Fomin, 1995. Slavery and Slave Trade among Banyang. *Paideuma* 41, 121–206.
Chickering, R., 1984. *We men who feel most German: a Cultural Study of the Pan-German League, 1886–1914*. Boston.
Chilver, E. M., 1961. Nineteenth-Century Trade in the Bamenda Grassfields, Southern Cameroons. *Africa und Übersee* 45, 233–258.
—— 1963. Native Administration in the West Central Cameroons 1902–1954, in Robinson, Kenneth & A.F. Madden (eds.), *Essays in Imperial Government: Presented to Margery Perham*. Oxford, 89–139.
—— 1966. *Zintgraff's Explorations in Bamenda, Adamawa and the Benue lands*. Buea.
—— 1967a. Paramountcy and Protection in the Cameroons: The Bali and the Germans, 1889–1913, in Gifford, P. und W. R. Louis (eds.), *Britain and Germany in Africa*. New Haven, 479–511.
—— 1967b. The Bangwa and the Germans: A Tail-piece. *Journal of the Historical Society of Nigeria* 4 (1), 155–60.
—— 1999. *Zintgraff's Explorations between the Coast and the Grassfields*. Mimeograph, by Friends of the Buea Archives, Oxford, unpublished.
Clarence-Smith, W.G., 1976. Slavery in Coastal Southern Angola. *Journal of South African Studies* 2(2), 214–33.
—— 1985. *The Third Portuguese Empire 1825–1975. A Study in Economic Imperialism*. Manchester.
—— 1991. The Hidden Costs of Labour on the Cocoa Plantations of Sao Tomé and Principe, 1875–1914. *Portuguese Studies* 6, 152–172.
—— 1993. Plantations versus Smallholder production of cocoa: the legacy of the German period in Cameroon, in Geschiere P. & P. Konings (eds.), *Itinéraires d'accumulation au Cameroun*. Paris & Leiden, 187–216.
Clarence-Smith, W. G. & François Ruf, 1996. Cocoa Pioneer Fronts: The Historical Determinants, in Clarence-Smith (ed.), *Cocoa Pioneer Fronts since 1800*. London & New York, 1–22.
Conradt, L., 1899–1900. Die landwirtschaftliche Regierungsstation Johann-Albrechts-Höhe. *Beiträge zur Kolonialpolitik* 1, 322–46.
—— 1900. Die Eingeborenen in der Umgebung der Station Johann-Albrechts-Höhe. *DKZ*, N.F. XIII (4), 33–36.
Conrau, G., 1894a. Aus Baliland. *DKB* 5, 190.
—— 1894b. Über das Gebiet zwischen Mundame und Baliburg. *Mitt. aus den deutschen Schutzgebieten* 7 (2), 99–104, 277–280.
—— 1898. Einige Beiträge über die Völker zwischen Mpundu und Bali. *Mitt. aus den deutschen Schutzgebieten* 11 (3), 194–202.
—— 1899. Im Lande der Bangwa. *Mitt. aus den deutschen Schutzgebieten* 12, 201–218, map.
Degener, 1912, 1955. *Wer ist Wer*, 6th & 12th editions, Berlin.
DeLancey, Mark W., 1978. Health and Disease on the Plantations of Cameroon 1884–1939, in G.W. Hartwig & K.D. Patterson (eds.), *Disease in African History*. Durham, NC, 153–79.
DeLancey, Mark W., with H. Bella Mokeba, 1990. *Historical Dictionary of the Republic of Cameroon*, London & Metuchen.

DeLancey, Virginia & M. DeLancey, 1972. *A bibliography of Cameroon Folklore.* Waltham, Mass.
Dieu, Michel & P. Renaud, 1983. *Atlas linguistique de l'Afrique Centrale*: *Le Cameroun. (Inventaire préliminaire).* Yaounde.
Dippold, Max, 1971. *Une Bibliographie du Cameroun: Les écrits en langue allemande.* Burgau & Yaounde.
Drechsler, Horst, 1996. *Südwestafrika unter deutscher Kolonialherrschaft. Die großen Land- und Minengesellschaften.* Stuttgart.
Drees, Ingenieur, 1894. Reise nach Batom. *DKB* 5, 45.
Duffy, James, 1967. *A question of Slavery.* London.
Dundas, R.W.M., 1922. *Assessment Report on the Bafaw,* Buea National Archives, bound typescript.
Eckert, Andreas, 1996. Cocoa Farming in Cameroon, c.1914 – c.1960: Land and Labour, in Clarence-Smith (ed.), *Cocoa Pioneer Fronts since 1800.* 137–153.
—— 1997 (2nd. edn.). *Die Duala und die Kolonialmächte.* Münster & Hamburg.
Egerton, F.C.C., 1957. *Angola in Perspective.* London.
Ejedepang-Koge, S. N., 1986. *The tradition of a people: Bakossi.* Yaounde.
Elango, Lovett Z., 1989. Trade and Diplomacy on the Cameroon coast in the nineteenth century, 1833–1879: the case of Bimbia, in M. Njeuma, (ed.), *Introduction to the History of Cameroon in the Nineteenth and Twentieth Century.* London, 32–62.
Eley, Geoff, 1980. *Reshaping the German Right. Radical Nationalism and Political Change after Bismarck.* New Haven & London.
Epale, S.J., 1985. *Plantations and Development in Western Cameroon, 1885–1975.* New York.
Epstein, K., 1959a. Matthias Erzberger and the German Colonial Scandals. *English Historical Review* 74, 639–663.
—— 1959b. *Matthias Erzberger and the dilemma of German Democracy.* Princeton.
Ernst, F., 1903. Die ersten Erfahrungen unserer Brüder in Bali. *Der evangelische Heidenbote* 10, 73–75 & 11, 86–87.
Erzberger, Matthias, 1906. *Die Kolonial-Bilanz. Bilder aus der deutschen Kolonialpolitik.* Berlin.
Esser, Anna (ed.), 1916. *Chronik der Familie Esser in Rheinland und Westfalen und deren verwandten Zweige.* Paderborn.
Esser, Max, 1897a. Meine Reise nach dem Kunene im nördlichen Grenzgebiet von Deutsch-Südwest-Afrika. *Verhandlungen der Gesellschaft für Erdkunde zu Berlin* 24, 103–113, map.
—— 1897b. Unsere Westafrikanischen Kolonien und ihr portugiesischer Nachbar, Supplement to the *DKZ* III, 6 February 1897, 9–15 & VIII, 27 March 1897, 33–36.
—— 1897c. Sitten der Hereros. *DKZ,* N.F. X (20), 193–194.
Evans, R. J., 1978. *Wilhelm II's Germany and the Historians.* New York.
Eyoh, Dickson, 1998. Through the prism of a local tragedy. *Africa* 68 (3), 338–359.
Eyzaguirre, Pablo B., 1988. Competing Systems of Land Tenure in an

African Plantation Society, in Downs, R. E. & S. P. Reyna (eds.), *Land and Society in Contemporary Africa*. Hanover, NH & London, 340–361.
Fabri, Friedrich, 1879. *Bedarf Deutschland der Colonien?* Gotha.
Fanso, V. G., 1989. Trade and Supremacy on the Cameroon Coast, 1879–1887, in M. Njeuma (ed.), *Introduction to the History of Cameroon in the Nineteenth and early Twentieth Century*. London, 63–87.
Fardon, Richard, 1983. A Chronology of Pre-Colonial Chamba History. *Paideuma* 29, 67–92.
—— 1988. *Raiders and Refugees: Trends in Chamba Political Development, 1750–1950*. Washington.
—— 1990. *Between God, the Dead and the Wild. Chamba Interpretations of Religion and Ritual*. Edinburgh.
—— 1996. The Person, Ethnicity and the Problems of 'Identity', in Fowler, I. & D. Zeitlyn (eds.), *African Crossroads*. Providence & Oxford, 17–44.
Fitzner, Rudolph, 1896. *Deutsches Kolonial-Handbuch*. Berlin.
Flegel, E. R., 1885. *Lose Blätter aus dem Tagebuche meiner Haussafreunde*. Hamburg.
Fobia IV, Chief, 1985. *Nguti and Environs*. Nguti (pamphlet).
Fohtung, M. G., 1992. Selfportrait of a Cameroonian; taken down by Peter Kalle Njie and edited by E. M. Chilver. *Paideuma* 38, 219–248.
Fomin, E. S. D. & V. J. Ngoh, 1998. *Slave Settlements in the Banyang Country 1800–1950*. Limbe: University of Buea Publications.
Förster, S., J. Mommsen & R.E. Robinson, 1988. *Bismarck, Europe and Africa*. Oxford.
Fowler, Ian & David Zeitlyn (eds.), 1996. *African Crossroads*. Providence & Oxford.
Friderici, G., 1898. Die Verhältnisse in Kamerun. *DKZ*, N.F.XI (18), 162–163.
Gardi, Bernard, 1994. *Kunst in Kamerun*. Museum für Völkerkunde und Schweizerisches Museum für Volkskunde. Basel.
Gausset, Quentin, 1997. *Les avatars de l'identité chez les Wawa et les Kwanja du Cameroun*. Doctoral Diss., Université Libre de Bruxelles.
Geary, Christraud, 1996. Political dress: German Style Millitary Attire and Colonial Politics in Bamum, in Fowler, I. & D. Zeitlyn (eds.), *African Crossroads*. Providence & Oxford, 165–192.
Geschiere, Peter, 1995. *Sorcellerie et politique en Afrique*. Paris.
Glauning, Hans, 1905. Bericht des Hauptmanns Glauning, Leiter der Station Bamenda, über seine Expedition nach Bali, Bameta und dem Südbezirk. *DKB* 16, 667–73.
—— 1906. Bericht des Hauptmanns Glauning über seine Reise in den Nordbezirk, 1905. *DKB* 17, 235–241.
Haas, Waltraud & Paul Jenkins, 1988. *Guide to the Basel Mission's Cameroon Archive*. Basel, multigraphed.
Halldén, E., 1968. *The Culture Policy of the Basel Mission in the Cameroons, 1886–1905*. Lund.
Harter, Pierre, 1981. Les perles de verre au Cameroun. *Arts d'Afrique noire* 40, 6–22.
—— 1986. *Arts Anciens du Cameroun*. Arnouville.

Hausen, Karin, 1970. *Deutsche Kolonialherrschaft in Afrika*. Zürich & Freiburg im Br.
Henderson, W.O., 1962. *Studies in German Colonial History*. London.
Hutter, Franz, 1891. Bericht über den Abschluss des Vertrages zwischen Dr. Zintgraff und Garega. *DKB* 2, 517–518, 541–544.
—— 1893a. Mein Aufenthalt bei den Balis von 1891–1893. *DKZ*, N.F. VI (8), 99–101.
—— 1893b. Allgemeiner Bericht über die Station Baliburg. *DKB* 4, 36–37, map.
—— 1899a. Der Abschluß von Blutsfreundschaft und Verträgen bei den Negern des Graslandes in Nordkamerun. *Globus* 74, 1–4.
—— 1899b. Politische und soziale Verhältnisse bei den Graslandstämmen Nordkameruns. *Globus* 74, 284–309.
—— 1900. Zur Lage in Kamerun. *Koloniale Zeitschrift* 9, 117–119.
—— 1902. *Wanderungen und Forschungen im Nord-Hinterland von Kamerun*. Braunschweig.
—— 1904. Völkergruppierung in Kamerun. *Globus* 86, 1–5.
—— 1905. Völkerbilder aus Kamerun. *Globus* 87, 234–238, 301–304, 365–370.
Ifeka, Caroline & E. Flower, 1999. Global Identities, Kinship and Witchcraft Trials in Boki Society. *JASO* 28(3), 311–338.
Ittmann, Johannes, 1955. Gottesvorstellungen und Gottesnamen im nördlichen Waldland von Kamerun. *Anthropos* 50, Sonderabdruck.241–264.
—— 1956. Von Totengebräuchen und Ahnenkult der Kosi in Kamerun. *Africa* 26 (4), 380–397.
—— 1960. Orakelwesen im Kameruner Waldland. *Anthropos* 55, 114–34.
Jaeck, H-P., 1960. Die deutsche Annexion, in Stoecker, H. (ed.), *Kamerun unter deutscher Kolonialherrschaft*. Vol. 1. Berlin. 29–95.
Jahoda, Gustav, 1998. *Images of Savages*. London.
James, Harold, 1989. *A German Identity*. London.
Jarausch, Konrad A., 1982. *Students, Society and Politics in Imperial Germany*. Princeton.
Jeffreys, M.D.W., 1957. The Bali of Bamenda. *African Studies* 16 (2), 108–113.
—— 1962a. Some Notes on the Customs of the Grassfields Bali of Northwestern Cameroons. *Afrika und Übersee,* 46 (3), 161–68.
—— 1962b. Traditional Sources prior to 1890 for the Grassfields Bali of Northwestern Cameroons. *Afrika und Übersee* 46 (3), 168–199; (4), 296–313.
Jenkins, Paul, 1981. Warum tragen die Missionäre Kostüme? *Historische Anthropologie* 4 (2), 292–302.
John, Hartmut, 1981. *Das Reserveoffizierkorps im deutschen Kaiserreich, 1890–1914*. Frankfurt & New York.
Johnston, H. H., 1885. The Portuguese possesions in West Africa. *Scottish Geographical Magazine* 1 (10), 465–482.
Kaberry, P. M. & E. M. Chilver, 1961. An Outline of the Traditional Political System of Bali-Nyonga, Southern Cameroons. *Africa* 31 (4), 355–71.
Keller, Werner, 1969. *The history of the Presbyterian Church in West*

Cameroon (with additional chapters by J. Schnellenbach & J. R. Brutsch). Victoria (Limbe).
Kemner, Wilhelm, 1922. *Was wir verloren haben. Aus der Geschichte der Westafrikanischen Pflanzungsgesellschaft* 'Victoria', Hamburg.
—— 1937. (1st edn.), 1941 (2nd edn.), *Kamerun.* Berlin.
Kitchen, Martin, 1968. *The German Officer Corps. 1890–1914.* Oxford.
Konings, Piet, 1993. *Labour Resistance in Cameroon.* Leiden, London & Yaounde.
Koschitzky, Max von, 1887. *Deutsche Colonialgeschichte.* Leipzig.
Krieger, Kurt, 1978. *Westafrikanische Plastik.* Vol. 1. Berlin.
Kuczynski, R. R., 1939. *The Cameroons and Togoland: a Demographic Study.* London.
Lagerberg, C.S.I.J & G.J. Wilms, 1974. *Profile of a Commercial Town in West-Cameroon: Research findings of a socio-anthropological enquiry in Kumba.* (A Study of Urban Problems in Africa). Tilburg.
Langheld, Wilhelm, 1909. *Zwanzig Jahre in deutschen Kolonien.* Berlin.
Latham, A. J., 1973. *Old Calabar, 1600–1891.* Oxford.
Lauffer, N., 1898a. Ein Auto da Fé in Kamerun, *Der evangelische Heidenbote*, 6.
—— 1898b. Der Riesengötze Dikoki aus Kamerun: Masken und Stäbe. *Der evangelische Heidenbote,* 47–48.
Leroy-Beaulieu, Paul, 1890. (1902, 5th.edn) *De la colonisation chez les peuples modernes.* Paris.
Lessner, Oberleutnant, 1904. Die Baluë- oder Rumpiberge und ihre Bewohner. *Globus* 86, 273–278, 337–344, 392–397.
Levin, Michael, 1976. *Family Structure in Bakossi: Social Change in an African Society.* Ph.D Diss. Princeton University.
—— 1977. Progress in the past: dilemmas of wealth in Bakosi, in Sandra Wallman (ed.), *Perceptions of Development.* London & New York, 78–86.
—— 1980. Export Crops and Peasantization: the Bakosi of Cameroun, in M. Klein (ed.), *Peasants in Africa: Historical and Contemporary Perspectives.* Beverly Hills & London, 221–241.
Louis, W.R., 1967. Great Britain and German Expansion in Africa, 1884–1919, in Gifford, P. & W.R. Louis (eds.), *Britain and Germany in Africa.* New Haven, 3–46.
Luschan, Felix von, 1897. *Beiträge zur Völkerkunde der deutschen Schutzgebiete.* Berlin.
Lutz, F., 1907. *Im Hinterland von Kamerun: über Bali nach Bamum.* Basel.
Mafiamba, P.C., 1966. The Ekot-Ngba. *Abbia* 14–15, 99–107.
Mansfeld, A., 1908. *Urwalddokumente.* Berlin.
—— 1909. Totemismus in Kamerun. *DKZ* XXVI (13), 218–219.
Mansfeld, A. & H. Reck, 1928. *Westafrika: Aus Urwald und Steppe zwischen Crossfluss und Benue,* with 143 plates and map. Munich.
Mbuagbaw, T., 1972. *History Bo Mbu* (Tali History). Multigraphed Pamphlet.
Martin, Rudolf,. 1913. *Jahrbuch der Millionäre.* Vols. 7 and 9. Berlin.
'Merkur', 1912. Kakaokultur der Eingeborenen in Kamerun. *Koloniale Rundschau* 5, 268–272.

Merz, Andreas, 1997. *Die Politik Bali-Nyongas gegenüber der Basler Mission und der deutschen Kolonialmacht*. Diss. Historical Seminar, University of Basel.
Migeod, Frederick W.H., 1924. British Cameroons: its tribes and natural features. *J. African Society* 23, 176–187.
—— 1925. *Through British Cameroons*. London.
Mohammadou, Eldridge, 1978. *Catalogue des archives coloniales allemands du Cameroun: Le service des Archives Nationales de Yaoundé*. Institute for the Study of Languages and Cultures of Asia and Africa, Tokyo.
—— 1990. *Traditions historiques des peuples du Cameroun central*, 2 Vols. Tokyo.
Moisel, M., 1903. Aus dem Schutzgebiete Kamerun. *Mitteilungen aus den deutschen Schutzgebieten* 16, 1–8.
—— 1904. Der Aufstand am Croß-Fluß. *DKZ*, N.F. XVII (10), 90–92, map.
—— 1908a. Eine Expedition in die Grasshochländer Mittel-Kameruns. Supplement to *DKZ* XXV (14, 15), April, 236–238, 267–272.
—— 1908b. Zur Geschichte von Bali und Bamum. *Globus* 93, 117–120.
Monga, Yvette, 1996. The Emergence of Duala Cocoa Planters under German Rule in Cameroon: A Case Study of Entrepreneurship, in Clarence-Smith (ed.), *Cocoa Pioneer Fronts since 1800*. London & New York. 119–136.
Morgen, Curt, 1893. *Durch Kamerun von Süd nach Nord: Reisen und Forschungen im Hinterlande 1889–1891*. Leipzig.
Moseley, L. H., 1899. Regions of the Benue. *Proceedings of the Royal Geographical Society* 14 (6), 630–37, map.
Mossner, Julius, 1935. *Adressbuch der Direktoren und Aufsichtsräte*. Berlin.
Ngo, V. Viban, 1987. *Survey and Mapping of Cameroon, 1884–1984*. 2 Vols., Vol. 2: a Cartobibliography. Ph.D. Thesis, London School of Economics.
Nicklin, Keith, 1996. 'Standing male figure'; Entry 5.4, in Tom Phillips (ed.), *Africa: the Art of a Continent*. (Royal Academy of Arts). London, 348.
Niger-Thomas, M., 1995. Women's Access to and the Control of Credit in Cameroon: The Mamfe Case, in *Money-Go-Rounds. The Importance of Rotating Savings and Credit Associations for Women*. S.Ardener, & S. Burman (eds.), Oxford & Washington, 95–110.
Njeuma, M. Z. (ed.), 1989. *Introduction to the History of Cameroon in the Nineteenth and early Twentieth Century*. London.
Nkwi, Paul N., 1989. *The German Presence in the Western Grassfields* (Research Report No. 37, African Studies Centre). Leiden, maps.
Northern, Tamara, 1984. *The Art of Cameroon*. Washington.
Nyamdi, Ndifontah B., 1988. *The Bali-Chamba of Cameroon*. Paris & Yaounde.
Oliver, Roland, 1957. *Sir Harry Johnston and the Scramble for Africa*. London.
O'Neil, Robert, 1987. *A History of Moghamo, 1865 to 1940. Authority and Change in a Cameroon Grassfields Culture*. Ph.D. Diss. Columbia University.
—— 1996. Imperialisms at the century's end: Moghamo Relations with Bali-Nyonga and Germany 1889–1908, in I. Fowler & D. Zeitlyn (eds.), *African Crossroads*. Providence & Oxford, 81–100.

Phillips, Anne, 1989. *The Enigma of Colonialism: British Policy in West Africa*. London & Bloomington.
Plehn, A., 1904. Beobachtungen in Kamerun. *Zeitschrift für Ethnologie* 6, 713–726.
Pogge von Strandmann, Hartmut, 1969. The domestic origins of Germany's colonial expansion under Bismarck. *Past and Present* 42, 149–159.
Pogge von Strandmann, Hartmut & Alison Smith, 1967. The German Empire in Africa and British Perspectives: a Historiographical Essay, in Gifford, P. & W. R. Louis, *Britain and Germany in Africa*. New Haven, 709–795.
Pohl, Hans (ed.), 1993. *Europäische Bankengeschichte*. Frankfurt.
Pradelles de Latour, C-H., 1995. The Initiation of the Dugi among the Péré. *JASO* 26 (1), 81–86.
Puttkamer, J. von, 1901. Expedition des Gouverneurs nach den Cross-Schnellen, *DKB* 12, 275–278.
—— 1912. *Gouverneursjahre in Kamerun*. Berlin.
Ramsay, H. von, 1901. Expedition des Generalbevollmächtigten der Gesellschaft Nordwest Kamerun. *DKB* 12, 234–238.
—— 1904. Meine letzten Reisen im Schutzgebiet Kamerun 1900–1904. *Mitt. des Vereins für Erdkunde zu Leipzig*, 27–30.
—— 1925. Entdeckungen in Nordwest-Kamerun, in Hans Zache (ed.), *Das deutsche Kolonialbuch*. Berlin.
Richards, Sarah C., 1998. *Dilemma of fertile virgins: women's sense of self in treatment choice in Bali*. Ph.D Dissertation, Boston University.
Rogozinski. Stefan von, 1884. Reisen im Kamerun-Gebiete. *Petermanns Mitteilungen aus Justus Perthes Geographischer Anstalt* 30, 132–139, map.
Rohde, Eckart, 1997. *Grundbesitz und Landkonflikte in Kamerun*. Münster & Hamburg.
Röschenthaler, Ute, 1993. *Die Kunst der Frauen. Zur Komplementarität von Nacktheit und Maskierung bei den Ejagham im Südwesten Kameruns*. Berlin.
—— 1998. Honoring Ejagham Women. *African Arts* 31 (2), 38–49, 92.
—— 1999a. Frauenbünde der Ejagham im Cross River Gebiet, In: *Afrikanische Frauen*. Coburg, 26–50.
—— 1999b. 'Fetische aus Bulo N'Guti'. Max Esser's 'Bakundu'-Skulpturen und der Handel im Waldland von Kamerun. *Tribus* 48, 147–171.
—— 1999c. Max Esser's 'Bakundu'-Fetishes. *African Arts* 32 (4), 76–80, 96.
—— 2000. Lokalität und Siedlungsgeschichte im Cross River Gebiet. *Zeitschrift für Ethnologie* 125.
Rudin, H.R., 1938. *Germans in the Cameroons, 1884–1914: A Case Study in Modern Imperialism*. New Haven.
Ruel, Malcolm J., 1969. *Leopards and Leaders: Constitutional Politics among a Cross River People*. London.
—— 1970. Were-animals and the introverted witch, in M. Douglas (ed.), *Witchcraft Confessions and Accusations*. London, 333–350.
Rüger, A., 1960a. Der Aufstand der Polizeisoldaten (Dezember 1893), in H. Stoecker (ed.), *Kamerun unter deutscher Kolonialherrschaft*. Vol. 1. Berlin, 97–147.

—— 1960b. Die Entstehung und Lage der Arbeiterklasse unter dem deutschen Kolonialregime in Kamerun (1895–1905), in Stoecker, H. (ed.), *Kamerun unter deutscher Kolonialherrschaft*. Vol. 1. Berlin, 149–242.

—— 1968. Die Duala und die Kolonialmacht 1884–1914, in H. Stoecker (ed.), *Kamerun unter deutscher Kolonialherrschaft*. Vol. 2. Berlin, 181–258.

Russell, A., 1958. The Kola of Nigeria and the Cameroons. *Tropical Agriculture* 32, 210–240.

Russell, S.W., 1980. *Aspects of Development in Rural Cameroon: Political transition amongst the Bali of Bamenda*. Ph.D. Diss. University of Pennsylvania.

Scheibler, H. & K. Wülfrath, 1939. *Westdeutsche Ahnentafeln*. Weimar.

Schilder, Kees, 1988. *State formation, religion and land tenure in Cameroon: a bibliographical study* (Research Report No. 32, African Studies Centre). Leiden.

Schlatter, Wilhelm, 1916. *Geschichte der Basler Mission*. Vol. 3. Basel.

Schlosser, Oberleutnant, 1904. Bericht über die Zustände im Crossbezirk. *DKB* 15, 735.

Schnee, Heinrich (ed.), 1920. *Deutsches Kolonial-Lexikon, 3 Vols.* Leipzig.

Schwartzkopf (Captain), 1898. Kamerun. *DKZ*, N.F. XI (5), 43.

Schwarz, Bernhard, 1886. *Kamerun: Reise in die Hinterlande der Kolonie*. Leipzig.

Seitz, Th., 1927/1929. *Vom Aufstieg und Niederbruch deutscher Kolonialmacht*, 2 Vols. Karlsruhe.

Semler, Johannes, 1905. *Togo und Kamerun: Eindrücke und Momentaufnahmen von einem deutschen Abgeordneten*. Leipzig.

Sharpe, Barrie, 1998. 'First the Forest': Conservation, 'Community' and 'Participation' in South-West Cameroon. *Africa* 68 (1), 25–45.

Simo, 1986. L'intelligentsia allemande et la question coloniale, in K. Ndumbe III (ed.), *Africa and Germany from Colonisation to Cooperation 1884–1986*. Yaounde, 181–202.

Singlemann (Konsul), 1912. Deutschlands Beziehungen zu Angola. *Koloniale Rundschau* 5, 272–280.

Skolaster, H., 1924. *Die Pallotiner in Kamerun*. Limburg a.d. Lahn.

Smith, Woodruff D., 1978. *The German Colonial Empire*. Chapel Hill.

—— 1986. *The ideological origins of Nazi Imperialism*. Oxford & New York.

Soénius, Ulrich, 1992. *Koloniale Begeisterung im Rheinland während des Kaiserreichs*. Köln.

Spellmayer, Hans, 1931. *Deutsche Kolonialpolitik im Reichstag*. Stuttgart.

Spengler, (Vizekonsul), 1894. Bericht des Kaiserlichen Vizekonsuls Spengler über die Anbaufähigkeit des Gebietes des Bezirksamts Victoria der Kolonie Kamerun. *DKB* 5, 282–288.

Staschewski, F., 1917. Die Banjangi. *Baessler-Archiv*, Beiheft 8.

Steimel, Robert, 1950. *Rheinisches Wappenlexikon*, vol. 2. Köln.

—— 1958. *Kölner Köpfe*. Köln.

Stengers, Jean, 1967. British and German Imperial Rivalry: A conclusion, in

Gifford, P. & W. R. Louis *Britain and Germany in Africa*. New Haven, 337–347.
Stetten, Max von, 1893. Das nördliche Hinterland von Kamerun. *DKB* 4, 33–35.
—— 1895. Bericht über seinen Marsch von Balinga nach Yola. *DKB* 6, 110, 135, 159, 180.
Stockhorst, Erich, 1967. *500 Köpfe. Wer war was im 3. Reich*, Berlin.
Stoecker, H. (ed.), 1960 . *Kamerun unter deutscher Kolonialherrschaft. Vol. 1*. Berlin.
—— 1968. *Kamerun unter deutscher Kolonialherrschaft* Vol. 2. Berlin.
—— 1977. *Drang nach Afrika*. Berlin.
—— 1986 . (English trans.) *German imperialism in Africa*. London & Atlantic Heights, N.J.
Stöckle, J., 1994. *Traditions, Tales and Proverbs of the Bali-Nyonga* (Archiv Afrikanistischer Manuskripte, Bd. 2). Köln.
Struck, B., 1909. Die Geheimbünde der Balong. *Globus*, 96, 123–128.
Sydow, E. von, 1954. *Afrikanische Plastik*, ed. G. Kutscher. Berlin.
Talbot, Percy Amaury, 1912. *In the Shadow of the Bush*. London.
Tardits, Claude (ed.), 1981. *Contribution de la Recherche Ethnologique à l'Histoire des Civilisations du Cameroun*, with maps, chronologies and figures, 2 Vols. Paris.
Tesch, Johannes, 1912. (and earlier editions) *Die Laufbahn der deutschen Kolonialbeamten*. Berlin.
Thiel, J.F., H-J. Koloss and others, 1986. *Was sind Fetische?* Museum für Völkerkunde. Frankfurt.
Thorbecke, F., 1911. Das Manenguba-Hochland. Sonder-Abdruck aus *Mitt. aus den deutschen Schutzgebieten*, 24 (5), 1–32, plates and map.
Thormählen, Johannes, 1884. Mittheilungen über Land und Leute in Kamerun. *Mittheilungen der geographischen Gesellschaft in Hamburg*. No.1, 328–334: also in *DKZ* 1884, I (21), 417–420.
Titanji, V. et al., 1988. *An Introduction to the Study of Bali-Nyonga (A Tribute to His Royal Highness Galega II, Traditional Ruler of Bali-Nyonga 1940–1985)*. Yaounde.
Turner, H. Ashby, 1967. Bismarck's Imperialist Venture: Anti-British in Origin? in P. Gifford & W. R. Louis (eds.), *Britain and Germany in Africa*. New Haven, 47–82.
Valdau, Georg, 1886. Eine Reise in das Gebiet nördlich vom Kamerungebirge. *Deutsche Geographische Blätter* (Bremen) 9, 30–48, 120–141.
—— 1890. Schilderungen aus Kamerun. *DKZ* III (9, 10, 12, 13, 14, 16), 108–109, 123–126, 146–149, 159–161, 171–172, 194–195, map.
Valentin, Peter, 1972. Plastiken der Kundu (Kamerun) im Basler Missionsmuseum. *Ethnologische Zeitschrift Zürich*, 2, 35–51.
—— 1974. Tanzschürzen der Bali-Frauen in Kamerun. *Ethnologische Zeitschift Zürich* 2, 185–195.
—— 1983. Cameroon: Do the Kundu-figures originate from the Balong? *Nachrichten/Newsletter, Basler Afrika Bibliographien* 7, (1) 2–4.
Van Slageren, J., 1972. *Les origines de l' Église Évangelique du Cameroun*. Yaounde & Leiden.

Vielhauer, D.A., 1936. Aus den Anfängen der Grasland Mission, in E. Kellerhals (ed.), *Fünfzig Jahre Basler Missionsarbeit*. Basel, 45–59.
Wagner, Hans, 1899a. *Etwas vom 'Afrikareisenden' Dr. Jur. Esser*. Berlin.
—— 1899b. Landconcessionen in unseren Colonien. *Der Weltmarkt* 13 (15), 299–301.
—— 1900. *Falsche Propheten. Gouverneur Liebert und seine Presse*. Berlin.
Warhurst, P.R., 1962. *Anglo-Portuguese Relations in South-Central Africa*. London.
Warnier, J-P., 1985. *Échanges, développement et hiérarchies dans le Bamenda pré-colonial (Cameroun)*. Stuttgart (Studien zur Kulturkunde 76).
—— 1995. Slave-trading without Slave-raiding in Cameroon. *Paideuma* 41, 251–272.
—— 1996. Rebellion, Defection and the Position of Male Cadets, in I. Fowler & D. Zeitlyn (eds.), *African Crossroads*. Providence & Oxford, 115–123.
Wehler, Hans-Ulrich, 1985. *The German Empire, 1871–1918*. Oxford & New York.
Wenzel, Georg, 1929. *Deutscher Wirtschaftsführer*. Berlin.
Wilhelm, H., 1981. Le commerce précolonial de l'ouest, in C. Tardits (ed.), *Contribution de la recherche ethnologique à l'histoire des civilisations du Cameroun*, Vol 2, 485–500, maps.
Willequet, J., 1967. Anglo-German Rivalry in Belgian and Portuguese Africa, in P. Gifford & W. R. Louis (eds.), *Britain and Germany in Africa*. New Haven, 245–273.
Winkler, H., 1960. Das Kameruner Proletariat 1906–1914, in H. Stoecker (ed.), *Kamerun unter deutscher Kolonialherrschaft*. Vol 1. Berlin, 243–286.
Wirz, A., 1972. *Vom Sklavenhandel zum Kolonialen Handel*. Zürich & Freiburg im Br.
Woermann, Eduard, 1926. Aus dem Tagebuch von Eduard Woermann, in Hans Zache (ed.), *Das deutsche Kolonialbuch*. Berlin, 261–263.
Wohltmann, Ferdinand, 1896. *Der Plantagenbau in Kamerun und seine Zukunft*. Berlin.
—— 1897. Die Ziele und Erfolge der deutschen Kolonialpolitik und die Bestrebungen der deutschen Kolonialgesellschaft. *DKZ* X, Supplement to 15, 61–64.
Wright, A. K. (ed.), 1958. *Victoria, Southern Cameroons, 1858–1958*. Victoria (Limbe).
Wyllie, I. A. (trans. & ed.), 1911. *Portuguese Planters and British Humanitarians: the case for S. Thomé*. Lisbon.
Zeitlyn, David, 1995. Eldridge Mohammadou on Tikar origins. *JASO* 26 (1), 99–104.
Ziemann, Grete, 1908. *'Mola Koko': Grüsse aus Kamerun*. Berlin.
Zintgraff, Eugen, 1892. Denkschrift des Dr. Zintgraff betreffend die Zukunft Kameruns. *DKB* 3, 104–108, 131–137.
—— 1895. *Nord-Kamerun*. Berlin.
Zöller, Hugo, 1885. *Forschungsreisen in der deutschen Colonie Kamerun*, 3 Vols. Berlin & Stuttgart.

Index

Abo, 140
Adametz, Hauptmann, 158, 162, 164
Aengeneyndt, Herr, 114
Ahn, Albert, 7, 16, 18, 28, 129, 167,
Akwa, 43–4
Ambaca, 118
Angola, 102, 114–18, 121–4, 168
　Boer presence, 118, 122, 128
　diamond finds, 126
　gold finds, 120, 122
　harbours, 65, 123
　labour regime, 31, 118, 119
　mineral prospects, 118
　railway prospects, 8, 10, 124
　water resources, 122, 126–7
Animal husbandry, 50, 52, 135, 141, 162
Ankermann, Bernhard, 109, 158, 179
Annobon, 40
Anti-semitism, 6, 11, 12, 167, 170
Anyang revolt, 149, 155
Arbitration courts, 45
Architecture, 72, 75, 80
Archives
　Basel Mission, 108, 157
　Buea National, 47, 155, 185
　German, 47–8, 101
　Yaounde National, 102, 159
Ardener, Edwin, 40, 106
Ardener, Shirley, 56, 155
Aschu, 78, 79

Ba Cubabe, 115, 122
Babessong (Ashong), 80, 109, 140, 157
Babum (Mabum), 105, 134
Babungo, 109
Baden-Baden, 15, 16, 21–2, 24–5
Baduma, 71, 104, 140, 183
Bafo, 75, 105, 134–5, 140
Bafreng (Nkwen), 86, 108
Bafut, 8, 108, 139, 147, 157, 175
Bagam, 164
Bailundu, 119
Bakoko, 50, 54
Bakum (Mbakem), 154
Bakuni, 79, 140
Bakundu, 56, 70, 72–4, 84, 104–6, 134, 140–1, 173, 178, 181–5
Bakweri, 21, 47, 56, 60, 63, 102, 107, 140, 151–2
Bali-Nyonga
　artefacts, 173, 175–8
　dress, 3, 85, 89, 109, 138, 176
　ethnic composition, 29, 101, 138
　German views on, 69, 70, 85, 94, 109, 133, 139, 159
　labour recruitment, 9, 13, 62, 65, 136, 139, 144, 147, 157–8, 163, 170

migrations, 88–9, 101
royal wives, 173, 179
scouts (*Bagwe*), 90, 110, 117, 177
slavery in, 84, 94, 108, 138
trade, 84, 103, 108, 135, 138
Treaty of 1891, 163
tributaries, 89, 110, 137, 156, 161, 164
Balong, 103, 134, 140–1, 181–4
Balue, 182
Balz, H., 106, 183
Bamenda Station and Division, 102, 147, 155, 158–63
　pacification, 155
　reduction in size, 162
Bamessong, 137, 157
Bamileke, 55, 109, 156, 159, 161–2
Bamum, 108–9, 156, 158, 162, 164
Banano, 119
Bande (Bandeng), See Mankon
Bandjoun, 156
Bangwa, 106, 148, 170
Banks
　Belgian, 10, 171
　German, 5–8, 12, 14, 17–20, 127, 171
　Luxemburg, 12
　Portuguese, 35
Banti, 137, 140, 187
Banyang, 76, 78, 106–7, 110, 134–7, 140–1, 147, 178
Baptists, *See* Missions
Barombi, 52, 56, 104–5, 134
Bascho, 147
Bassus, Thomas Baron de, 12
Bastion, Adolph, 104
Batanga, 50–1
Batum (Batom), 75, 105
Beads, 59, 76–8, 83, 86–9, 91, 107–8, 135, 137–8, 144
Bell, King, 43, 44, 67, 153
Bell, Prince Manga, 67
Bell, Rudolph Manga, 153
Bengo, 118
Benguella, 117, 119, 121, 126
Berliner Blatt, 167, 171
Berlin Geographical Society, 11, 167, 171
Besser, Hauptmann, von, 58, 63, 110, 148, 154, 171
Bibundi, 10, 48, 53, 57–8, 151
Bimbia, 47, 53, 59, 68, 100
　chief of, 44
Böckner, G., 103, 141
Boer, 118, 122, 124, 126
Bohner, Missionary, 184
Boki, 106, 147
Bombe Mission, 181, 185
Bonge, 53, 56, 181, 183

Index

Botanic Gardens, 52, 58, 80, 103
Buchner, Max, 47, 182–3, 185
Buea, 47, 52, 56, 60, 102, 140, 153, 155, 157, 161, 185
Bulo Nguti, 72, 140, 173, 180–1, 185
Burton, R.F., 63

Cabinda, 35, 114, 119, 126
Camwood, 85, 134
Cannibalism, 73, 78, 105
Capello, Guilhelmo de, 120
Cape Verde Islands, 31–2, 126
Caro, Georg von, 18, 28
Carvings, *See* Sculpture
Cassinga, 122
Catumbella, 118
Cloth, 36, 50, 59, 76, 78, 87, 90, 135, 137–8, 144, 175, 178
Cocoa, 3, 9, 13–14, 31, 35–41, 51–3, 55, 58, 61–3, 102, 149
   Brown Rot disease, 13, 14
Coffee, 31, 36, 38–40, 51–3, 102–3, 118–9
Cologne, 3–7, 12, 14, 16–18, 20, 90
Colonial administration, 11, 154
Colonial scandals, 154, 169
Colonial societies
   DKG, 7, 11, 63
Concession companies,
   GNK, 19, 103, 108, 143–8, 154, 169, 171
   GSK, 5, 10, 148
Conradt, L., 104, 181, 183
Conrau, G., 18, 47, 104–5, 107, 140, 148, 170
Conte de Valle Flor, 36
Coroca River, 119, 121
Cross River, 48, 79, 105, 143, 146, 159
Cubango, 118
Cult agencies, *See* Secret Societies
Cunene, 117, 119, 120–27, 167, 169
Cung, Jan, 33

Dahomeyans, 54, 56, 103, 109
Damara, 119
Dapper, 43
Deichmann, W.L., 17, 19
Dernburg, Friedrich, 63
Diehl, A. 103, 147, 154
Difang, 136, 140–1
District administration, 45, 68, 102, 106, 141, 147, 152–3, 161–3
Dogs, 39, 73–4, 84, 95
Douala (Duala), 13, 43, 44, 47, 50–2, 54–6, 59, 60, 62–3, 103, 109, 113, 135, 174–5, 183–5
Douglas, Sholto, von, 6, 10, 18, 102, 171
Drum language, 59, 69
Duelling, 11, 168, 171
Dundas, R.W.M., 185
Dunstan, Elisabeth, 148
Dutch West Indies Co., 32
Dutreux, Antoine Auguste ('Tony'), 12
Dutreux, Emma Eleonore Flora Thérèse, 12
Dutreux, Franziska Elisabeth (Lily), 12

Ebermaier, Karl, 163
Edea, 67
Ejagham, 106–7, 141, 146, 154, 184
Ekoi, 148
Ernst, Missionary, 159
Erzberger, Matthias, 154

Esser, Adele Josephine, 6, 7, 16
Esser, Elizabeth, 28
Esser, Ferdinand Joseph, 5
Esser, Max
   The Esser 'Affair', 11, 12, 117, 167
   board memberships, 14–16, 18, 20–1
   correspondence with Graf von Linden, 174–5, 177
   education, 4, 5
   family history, 4–7, 18
   honours, 19, 167, 171, 176, 185
   marriage, 12, 19
   military service, 3, 15, 16
   route map, 188–9
   views
      on African cleanliness, 114, 115
      on Bakweri, 102
      on Bali characteristics, 133, 175
      on Duala, 113, 114
      on women and fashion, 77, 178
   wealth, 13
Esser, Robert Joseph, 4–7, 15–18, 24, 28
Ethnic composition, 101
   as described by Esser, 54, 173
   as described by Migeod, 105, 140
   as described by von Stetten, 105, 133–8, 140, 178
   later sources for SW Province, 106
European health, 51, 57, 60, 114

Fabri, Friedrich, 7
Fé, Princess, 109
Fernando Po, 36, 40, 43, 63, 65
Fitzner, R., 55
Fohtung, Max, 106, 110, 148
Fonte, 107, 137
Fonyonga, 107–8, 157, 159–63, 165
Fournier-Baudach, Walter, 10, 12, 114, 121, 126, 128, 168
Friderici, Herr, 53, 59, 63
Fulani, 151

Gabon, 43
Gardi, B., 181, 183
Garega (Galega), 10, 65, 69, 79, 81, 82–4, 86, 87– 98, 100–1, 113, 137, 144–6, 174, 176–8
Gawolbe, 159
Gemmingen, Freiherr von, 174, 176
German emigration, 101, 111
German political groups
   Centre Party, 128, 154
   National Liberals, 129
   Progress Party, 55
   Pan-German League, 171
   Social Democrats, 128
German South-West Africa, 8, 31, 63, 101, 117, 120–1, 124–6
Glauning, Captain Hans, 147, 155–9, 161, 164
Grassfields, 79–81, 89
Great Fredericksburg, 32–3
Guassi, 85–6, 89, 176–7
Guns, 8, 45, 76, 82, 92–3, 98–9, 104, 135

Hanno, 43
Hansemann, Adolf, von, 8, 17
Hartmann, von, 123
Hausa, 54, 82, 87–8, 107, 137, 139, 151, 174, 178
Herero, 19, 122

INDEX 203

Herodotus, 43
Hewett, Consul, 44–5, 47
Hiller, Max, 18, 143
Hoesch, Hermann, 15, 20
Hoesch, Victor, 8, 12, 13, 15, 18–20, 31, 130
Hohenlohe-Öhringen, Prince, 19, 128, 143
Holy Ghost Fathers, 118
Hottentots, 119–20
Huilla, 118
Humbe, 119, 122, 128, 168
Hutter, Lt. Franz, 83, 104–11, 133, 159, 163, 178

Ikiliwindi, 70–2, 104, 134, 140–1, 173, 176–7, 181
Islam, 88–9, 109
Isongo Udje, 53
Ivory, 44, 50–1, 61, 82, 84, 91, 93, 98–9, 118, 122, 135–6, 138, 144, 148, 158, 177–8

Johann-Albrechtshöhe, 52, 68
Johnston, Sir Harry, 40
Jordan, Christian, 12

Kaufmann-Asser, Jakob, 6
Kayser, Dr, 103, 144–5
Kemner, Wilhelm, 3, 8, 10, 13, 14, 20–1, 34, 41, 55, 63, 110, 130, 165, 169, 172
Kionka, Herr, 110, 144, 146
Knobloch, Lt. von, 159
Knorr, Admiral, 44–5
Knutson, Herr, 53, 55–6, 104
Kola, 79, 80–1, 83–4, 95, 99, 107, 145–6, 157, 178
Kolonialblatt, Deutsches, 9, 133, 167
Koloniale Zeitschrift, Deutsche, 83, 168, 170–1
Kolonialzeitung, 19, 167
Kom, 108, 156, 162
Kombone, 74–5, 105, 134, 140–1
Koschitsky, Max von, 33, 47
Kosi (Bakossi), 55, 104, 106, 140
Kru, 53, 60, 69
Kumba, 106, 134, 140, 183
Kuti Agricultural Station, 162

Labour Supply
  competition for, 4, 147
  forced labour, 4, 9, 87, 147, 161–2
  health, 13, 62, 151, 170
  Liberian, 53–4, 60, 113, 150
  mortality, 13, 64, 164, 170
  penal labour, 56, 63, 147, 152, 154, 163
  police and armed forces, 9, 147
  recruitment policy, 9, 20, 47, 62, 64, 69, 144, 146–8, 150, 159, 162–3, 170
  women as labourers, 54, 56, 63, 138
Land policy
  Crown land, 46, 143
  registration, 46, 113, 144
  unoccupied land, 46, 48
Langen, Eugen, 8
Langheld, Wilhelm, 13
Language distribution, 106, 134, 159, 189–90
Lauffer, Missionary, 181, 183, 185
Leroy-Beaulieu, Professor, 127
Lessner, Lt., 48, 104, 182
Leuschner, Herr, 52
Leutwein, Major, 119
Limbe (Victoria), 3, 9, 10, 13, 45, 47, 52, 58, 60, 65, 66, 69, 80, 100, 106–7, 110, 113, 123, 126, 139, 144–5, 151, 164, 170

Livestock. *See* Animal husbandry
Loanda, 114, 117–20, 125
Loo, W. van der, 149, 163
Löwenstein, Prince Alfred zu, 14, 18–19
Luschan, Felix, von, 173, 178

Makua Matafe, 121
Mankon, 108, 147
Man o' War Bay, 45, 53
Mansfeld, Dr Alfred, 48, 105, 107–8, 110, 154
Mbo-Miyenge, 134
Mendonca, J.H.M., 36
Menzel, Lt., 155–60
Migeod, F.W.H., 105, 140
Missions
  American Presbyterian, 55
  Baptist, 43, 45, 47, 63
  Basel, 101–5, 110, 147, 152, 156–9, 184
  Pallotine, 102
Miyimbi, 74, 78, 96, 136–7, 187
Moghamo, 108, 161
Moisel, Max, 109, 154, 171
Mokundange, 53
Mokurru, chief, 71
Mondoleh Island, 58–9
Morgen, Captain Curt von, 54
Mortuary rites, 115
Mossamedes, 8, 69, 117–21, 123, 125–6, 168–9, 171
Mt Cameroon, 9, 35, 40, 44, 49, 51–3, 60, 62, 76, 105, 182
Mundame, 49, 67–71, 76, 99, 103, 134–5, 139–40
Mundimba, 122
Mungo River, 49, 50, 67–8, 74, 99, 103, 134–5, 152
Muquich, 121
Museums
  Barbier-Müller Collection, Geneva, 181
  Basel Mission Museum, 181
  Berlin Museum für Volkerkunde, 72, 159, 173, 178, 181, 183,
  Linden Museum, Stuttgart, 173–4, 176, 178, 180, 185
Muyuka, 184
Myi Buba, 122

Nachtigal, Dr, 44, 113
Nachtigal, The, 57, 58, 114, 126, 144
Ngola, 105, 182–4
Nguti, 105–6, 134–7, 140–1, 185
Nicklin, Keith, 181, 185
Njoya, 156
Nkongsamba, 164
Novo Redondo, 65, 119
Nssakpe, 146–7
Nso', 147–8, 156, 160, 162, 164

Oechelhäuser, Wilhelm, 8, 13, 19
Old Calabar, 43
Ossidinge, 147, 162
Ovambo, 122

Palm oil, 44, 51, 66, 79, 152–61
Palmas, 53
Pavel, Lt. Colonel von, 147, 157, 159
Petroleum exploration, 13
Picht, H.F., 15, 20–1
Plantation companies
  Debundscha, 13, 15, 20, 54

KLPG, 9, 19, 53, 59, 63, 170
Meanja, 13–15, 20
WAPB, 10
WAPV, 3, 10, 13, 14, 16, 18, 21, 41, 47, 63, 102, 109–10, 130, 143–4, 146–50, 158, 162–5, 167, 169–70
Plantations repurchased in 1924, 21
Plehn, A. Dr, 105
Port Alexander, 121
Potatoes, introduction of, 84, 134
Preuss, Dr, 47, 52, 58, 60–1, 68, 110, 126
Principe, 31, 35–40, 65, 117, 126
Prussia, 5, 6, 13, 17, 20, 33, 46, 153, 172
Pückler-Limpurg, Graf von, 147
Punitive expeditions, 47, 100, 146, 148, 155, 158–9, 161, 170
Puttkamer, Jesco von, 3, 9, 13, 45, 47–8, 54, 63, 102, 144–5, 154, 158, 169, 185

Queiss, Lt., 47, 146, 148

Raber, Max, 167
Rackow, Hermann, 58, 170
Railroads, 153, 162
Ramsay, Hans von, 107–8, 146–8, 171
Raphia vinifera, 70, 72, 79
Ratzel, Friedrich, 168
Rehbein, Herr, 53, 100
Ribeiro, Dr Matheus Augusto, 35
Richthofen, Freiherr von, 18
Rio Campo, 45
Rio del Rey, 45, 47, 49, 56, 134
River crossings, 66, 71, 74–6, 94, 97, 99,
Roman Catholicism in Sao Thomé, 36
Rubber, 113–15, 20–1, 32, 38, 51–2, 61, 76, 118, 125, 134–7, 144, 147–8, 157

Sabi, 78, 94, 96, 136, 140
Sampaio Nunes, 121
Sao Thomé, 9, 31, 35–41, 57, 60, 65, 102, 114, 117, 119, 126
Scharlach, Julius, 6, 10, 19, 111, 171–2
Schlatter, W., 47, 110, 185
Schöller, Max, 18, 143, 171
Schroeder, Dr Ernst, 7, 14
Schwarz, Dr, 70, 104, 107
Sculpture, 21, 72, 104, 107, 173, 175, 177, 181–3
Secret Societies, 73, 104–5, 181, 183–5
Seitz, Theodor, 3, 18, 48, 59, 63, 113–4, 144
Semler, Dr Johannes, 149, 151–4
Serpa Pinto, 31
Slavery, 32, 35, 39, 60, 65, 73–4, 78, 84, 87, 92, 94, 102, 105–7, 110, 115, 119, 122, 138, 141, 150–1, 183
  internal slave trade, 74
  slave sacrifice at chief's deaths, 31, 73–4, 78, 83, 87
  slave settlements, 103, 137, 141
Spengler, Herr, 9, 10, 38–40, 60, 62, 102, 114, 126, 169
Soden, Freiherr Julius von, 18, 45, 47, 56, 102, 113
Sommerfeld, Lt von, 161
Steinäcker, von, 134
Stetten, Max, von, 103, 105–7, 133–4, 137–41, 178
Strümpell, Lt, 147

Sugar, 8, 31, 36, 39, 50, 52, 68, 87–8, 118–19, 121
Szolc-Rogozinski, 45, 182

Tägliche Rundschau, 11, 19, 128, 167, 170–2
Talbot, P.A., 108
Tayo, 96–100, 110, 140
Technology transfer, 14
Theuss, Herr, 53
Thorbecke, F., 141
Tiger Bay, 8, 120, 123, 125, 169
Tikar, 140, 189
Tinto, 76, 106, 135–6, 139–40, 147, 163
Tobacco, 9, 31, 39, 51–2, 53, 61, 76–7, 82, 107, 119, 135, 137, 162
  pipes, 82, 173–8
  snuff, 77, 106–7, 135
Togo, 55, 101–2, 127, 149, 151, 164, 176
Trade
  barter rates, 60, 70, 113
  in old uniforms, 144
  middlemen, 59, 103, 134, 138, 143
  with English factories, 76, 78–9, 137
Traders
  Ambas Bay Trading Company, 111
  Bristol and Liverpool, 43
  Firma C. Woermann, 8, 9, 43–4, 47, 111
  Jantzen & Thormählen, 43, 67, 103, 134, 137
  John Holt, 148
  Swedish, 53, 55–7, 104–5, 118, 182
Tropenpflanzer, Der, 55, 111, 170

Uandi, daughter of Ashong chief, 109

Vai, 47, 62, 69–70, 81, 86, 89–90, 121, 139
Valdau, G., 55–6, 105, 182
Vielhauer, D.A., Missionary, 109–10, 164
Vietor, J.R., 102
Voss, Thomas, 47

Wagner, Dr Hans, 11, 12, 19, 128, 168–171
Widekum, 156
Wilhelm II, Emperor, 10, 11, 19, 20, 38, 88, 167–8
Woermann Shipping Line, 66–7
Wohltmann, Professor F., 12, 19, 51, 54–6, 111, 127–8, 171
Women
  as booty to soldiers, 157, 160
  as diplomatic gifts, 95, 109
  as penal labour, 56, 63
World War, First, 15, 21
World War, Second , 5, 17, 20–1, 172, 178
Wuri, 49, 50, 151
Württemberg, King of, 174, 185
Wute, 149, 151, 189

Zentral-Dombau-Verein, 5–7, 17
Ziemann, Dr, 109, 150
Zimmerer, Eugen von, 9
Zintgraff, Eugen, 8–10, 18, 31, 39, 52, 54, 56, 63, 65, 67, 69, 71, 82–3, 90, 95, 100, 102–10, 114, 121, 126, 133, 137–41, 144–6, 154, 157, 163, 169–71, 173, 178, 183
Zöller, Hugo, 7, 47, 56, 115

www.ingramcontent.com/pod-product-compliance
Lightning Source LLC
Chambersburg PA
CBHW071158070526
44584CB00019B/2836